ADVANCED SUBSIDIARY

Revise AS
Biology

Author

John Parker

Contents

Specification lists

AQA A Biology

MODULE	SPECIFICATION TOPIC	CHAPTER REFERENCE	STUDIED IN CLASS	REVISED	PRACTICE QUESTIONS
Module 1 (M1)	Biological molecules	1.1, 1.2, 1.3, 1.4			
	Biochemical tests	1.5			
	Microscopy	2.2			
	Cell ultrastructure	2.1			
	Cell fractionation	2.2			
	Transport across cell membranes	4.1, 4.2			
	Enzymes	3.1			
	Inhibitors	3.2			
	Tissues and organs	2.3			
	Heart and circulation	5.3, 5.4			
	Effects of exercise	9.1			
Module 2 (M2)	Enzyme applications	3.4			
	Cell division	6.2			
	DNA	6.1, 6.2			
	Protein synthesis	6.1			
	Gene technology	6.3			
	Immunology	9.3			
	Electrophoresis	6.3			
	Genetic fingerprinting	6.3			
	Isolating genes	6.3			
	Ethical issues	6.4			
	Adaptations to environment	4.3, 5.6			
	Commercial crops	8.2			
	Fertilisers and pesticides	8.2			

Examination analysis

The specification comprises three compulsory modules. In modular tests 1 and 2, all questions are compulsory; they consist of structured questions and questions requiring extended answers.

Module 1	1 hour 30 minutes examination	35%
Module 2	1 hour 30 minutes examination	35%
Module 3	Centre-assessed coursework	30%

Note: the Centre-assessed coursework tests 10 different skills. Each may be assessed several times during the course Only the best mark in each of the ten skills counts. Marks awarded by your teacher are subject to change by an external moderator.

AQA B Biology

MODULE	SPECIFICATION TOPIC	CHAPTER REFERENCE	STUDIED IN CLASS	REVISED	PRACTICE QUESTIONS
Module 1 (M1)	Biological molecules	1.1, 1.2, 1.3, 1.4, 1.6			
	Biochemical tests	1.5			
	Chromatography	1.5			
	Cell ultrastructure	2.1			
	Microscopy	2.2			
	Tissues and organs	2.3			
	Transport across cell membranes	1.3, 4.2			
	Exchange with environment	4.1			
	Enzymes	3.1, 3.4			
	Inhibitors	3.2			
	Digestion	3.3			
Module 2 (M2)	Genes and DNA	6.1			
	The genetic code	6.1			
	Protein synthesis	6.1			
	Mutation	6.1			
	Cell division	6.2, 7.1			
	Gene technology	6.3			
	Industrial fermenters	6.3			
	Genetically modified organisms	6.4			
	Ethical issues	6.4			
Module 3(a) (M3)	Mass transport	5.1			
	Heart and circulation	5.2, 5.3, 5.4			
	Oxygen transport	5.4			
	Ventilation	4.3			
	Nervous control of heart	5.2			
	Transport in plants	5.5			
	Xylem and phloem	5.5, 5.6			
	Transpiration	5.6			
	Xerophytes	5.6			
	Translocation	5.6			

Examination analysis

The specification comprises three compulsory modules. In modular tests 1, 2 and 3(a) all questions are compulsory; they consist of structured questions and questions requiring extended answers.

Module 1	1 hour 15 minutes examination	30%	**Module 3(a)**	1 hour 15 minutes examination	25%
Module 2	1 hour 15 minutes examination	30%	**Module 3(b)**	Centre-assessed coursework	15%

OCR Biology

MODULE	SPECIFICATION TOPIC	CHAPTER REFERENCE	STUDIED IN CLASS	REVISED	PRACTICE QUESTIONS
Module 2801 (M1)	Cell ultrastructure	2.1			
	Microscopy	2.2			
	Tissues and organs	2.3			
	Biological molecules	1.1, 1.2, 1.3, 1.4			
	Biochemical tests	1.5			
	Enzymes	3.1			
	Inhibitors	3.2			
	Cell membranes	4.1, 4.2, 4.3			
	Transport across cell membranes	1.3, 4.2			
	DNA	6.1			
	Protein synthesis	6.1			
	Cell division	6.2			
	Ecosystems	8.1			
	Energy transfer	8.2			
	Nitrogen cycle	8.3			
Module 2802 (M2)	Diet	9.1			
	Gaseous exchange in lungs	4.3			
	Effects of exercise	9.1			
	Smoking and disease	9.1			
	Infectious disease	9.2			
	Immunity	9.3			
	Disease control	9.3			
Module 2803 (M3)	Transport in mammals	5.4			
	Blood	5,4, 9.3			
	Heart and circulation	5.2, 5.3, 5.4			
	Nervous control of heart	5.2			
	Transport in plants	5.5			
	Xylem and phloem	5.5, 5.6			
	Water potential	4.2			
	Translocation	5.6			
	Transpiration	5.6			
	Xerophytes	5.6			

Examination analysis

The specification comprises three compulsory modules. In module tests all questions are compulsory; they consist of structured questions and questions requiring extended answers.

Module 2801 1 hour 30 minutes examination 30%	**Module 2803** 1 hour examination	20%
Module 2802 1 hour 30 minutes examination 30%	Centre-assessed coursework **or**	20%
	1 hour 30 minute practical examination	20%

Edexcel Biology

MODULE	SPECIFICATION TOPIC	CHAPTER REFERENCE	STUDIED IN CLASS	REVISED	PRACTICE QUESTIONS
Unit 1 (M1)	Biological molecules	1.1, 1.2, 1.3, 1.4			
	DNA	6.1, 6.2			
	The genetic code	6.1			
	Protein synthesis	6.1			
	Enzymes	3.1			
	Immobilised enzymes	3.4			
	Enzyme applications	3.4			
	Cell ultrastructure	2.1			
	Microscopy	2.2			
	Transport across cell membranes	1.3, 4.2			
	Water potential	4.2			
	Tissues and organs	2.3			
	Cell division	6.2			
Unit 2B (M2)	Exchange surfaces in plants and animals	4.1, 4.2, 4.3			
	Gaseous exchange in lungs	4.3			
	Digestion and absorption	3.3			
	Transport in animals	5.2, 5.3, 5.4			
	Transport in plants	5.5			
	Xylem and phloem	5.5, 5.6			
	Transpiration	5.6			
	Translocation	5.6			
	Heart and circulation	5.2, 5.3, 5.4			
	Nervous control of heart	5.2			
	Blood	5.4, 9.3			
	Haemoglobin; dissociation curves	5.4			
	Adaptations to environment	4.3, 5.6			
	Sexual reproduction in plants	7.2			
	Sexual reproduction in humans	7.3			
	Oogenesis and spermatogenesis	7.3			
	Menstrual cycle	7.3			
	Birth and lactation	7.3			
Unit 3 part (a) (M3)	Energy flow though an ecosystem	8.1			
	Food chains and webs	8.1			
	The nitrogen cycle	8.3			
	The carbon cycle	8.3			
	Effects of human activities on the environment	8.5			

Examination analysis

The specification comprises three compulsory modules. In module tests all questions are compulsory; they consist of structured questions and questions requiring extended answers.

Unit 1	1 hour 30 minutes examination	33.3%		**Unit 3 part (a)**	1 hour examination	18.1%
Unit 2B	1 hour 30 minutes examination	33.3%		**Unit 3 part (a)**	Centre-assessed coursework	15.2%

WJEC Biology

MODULE	SPECIFICATION TOPIC	CHAPTER REFERENCE	STUDIED IN CLASS	REVISED	PRACTICE QUESTIONS
Unit 1 (M1)	Biological molecules	1.1, 1.2, 1.3, 1.4			
	Biochemical tests	1.5			
	Cell ultrastructure	2.1			
	Microscopy	2.2			
	Transport across cell membranes	4.1, 4.2			
	DNA	6.1, 6.2			
	Protein synthesis	6.1			
	Cell division	6.2			
	Enzymes	3.1			
	Inhibitors	3.2			
	Immobilised enzymes	3.4			
	Enzyme applications	3.4			
	Blood	5.4, 9.3			
	Haemoglobin; dissociation curves	5.4			
	Ecosystems and energy transfer	8.1			
	Food chains and webs	8.1			
	Predators and prey	8.1			
	Colonisation and succession	8.4			
	Effects of human activities on the environment	8.5			
	Pest control	8.2			
	Mass transport	5.1			
Unit 2 (M2)	Heart and circulation	5.2, 5.3, 5.4			
	Exchange with environment	4.2, 4.3			
	Root structure and function	5.5			
	Transpiration	5.6			
	Xerophytes	5.6			
	Translocation	5.6			

Examination analysis

The specification comprises three compulsory modules. In module tests all questions are compulsory; they consist of structured questions and questions requiring extended answers.

Unit 1   1 hour 40 minutes examination   35%

Unit 2   1 hour 40 minutes examination   35%

Unit 3   3 hours 45 minutes of practical work in centres, assessed by external assessor   30%

NICCEA Biology

MODULE	SPECIFICATION TOPIC	CHAPTER REFERENCE	STUDIED IN CLASS	REVISED	PRACTICE QUESTIONS
Module 1 (M1)	Biological molecules	1.1, 1.2, 1.3, 1.4			
	DNA	6.1, 6.2			
	Protein synthesis	6.1			
	The genetic code	6.1			
	Gene technology	6.3			
	Enzymes	3.1			
	Immobilised enzymes	3.4			
	Enzyme applications	3.4			
	Cell ultrastructure	2.1			
	Microscopy	2.2			
	Transport across cell membranes	4.1, 4.2			
	Water potential	4.2			
	Tissues and organs	2.3			
Module 2 (M2)	Photosynthesis				
	Energy flow though an ecosystem	8.1			
	Food chains and webs	8.3			
	The nitrogen cycle	8.3			
	The carbon cycle	8.5			
	Effects of human activities on the environment	8.5			
	Cell division	6.2			

Examination analysis

The specification comprises three compulsory modules. In module tests all questions are compulsory; they consist of structured questions and questions requiring extended answers.

Module 1	1 hour 30 minutes examination	40%	Centre-assessed coursework	20%
Module 2	1 hour 30 minutes examination	40%		

AS/A2 Level Biology courses

AS and A2

All Biology A Level courses being studied from September 2000 are in two parts, with a number of separate modules or units in each part. Most students will start by studying the AS (Advanced Subsidiary) course. Some will go on to study the second part of the A Level course, called A2. It is also possible to study the full A Level course in either order. Advanced Subsidiary is assessed at the standard expected halfway through an A Level course i.e. between GCSE and A Level. This means that the new AS and A2 courses are designed so that difficulty steadily increases:

- AS Biology builds from GCSE Science/Biology
- A2 Biology builds from AS Biology.

How will you be tested?

Assessment units

AS Biology comprises three units or modules. The first two units are assessed by examinations. The third component usually involves some method of practical assessment (this is dependent on the Examination Group). Examination Groups use *either* centre-assessed coursework *or* a practical examination.

Centre-based coursework involves practical skills marked by your teacher. The marks can be adjusted by moderators appointed by the awarding body.

If a practical examination is an option, it is based on identical skills to the Centre-assessed option. Some groups also include another part to the third component. This is a short examination of further content.

For AS Biology, you will be tested by three assessment units. For the full A Level in Biology, you will take a further three units. AS Biology forms 50% of the assessment weighting for the full A Level.

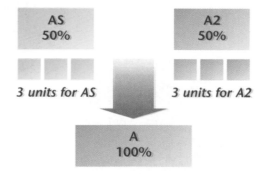

Tests are taken at two specific times of the year, January and June. It can be an advantage to you to take a unit test at the earlier optional time because you can re-sit the test, **(only once!)**. The best mark from the two will be credited and the lower mark ignored.

Each unit can normally be taken in either January or June. Alternatively, you can study the whole course before taking any of the unit tests. There is a lot of flexibility about when exams can be taken and the diagram below shows just some of the ways that the assessment units may be taken for AS and A Level Biology.

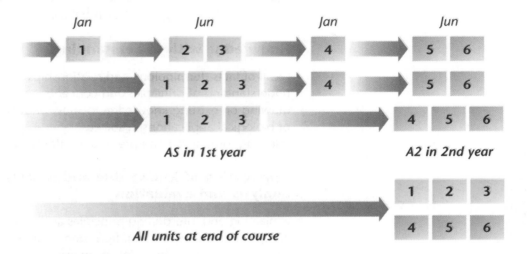

AS in 1st year   **A2 in 2nd year**

All units at end of course

If you are disappointed with a module result, you can resit each module once. You will need to be very careful about when you take up a resit opportunity because you will have only one chance to improve your mark. The higher mark counts.

A2 and synoptic assessment

Many students who have studied at AS Level may decide to go on to study A2. There are three further units or modules to be studied. Some units are optional, so it is the choice of the Centre e.g. a biotechnology unit may be chosen, or one of an ecological nature. Every A Level specification includes a 'synoptic' assessment at the end of A2. Synoptic questions make use of concepts from earlier units, bringing them together in holistic contexts. Examiners will test your ability to inter-relate topics through the complete course from AS to A2.

Coursework

Coursework may form part of your A Level Biology course, depending on which specification you study. Where students have to undertake coursework, it is usually for the assessment of practical skills but this is not always the case.

Key skills

These are new! Your work in Biology AS and A2 can be used to gain a further award, the key skills qualification. This helps you to develop important skills that are needed, whatever you do beyond A Level. The key skills include: Application of number, Communication and Information technology. There are three levels of award (1–3). Biology AS and A2 students have opportunities to study one or more of the key skills. You must collect evidence together in a portfolio to show your level of competence. The awarding body specification will show opportunities of appropriate topics which can also be used to develop key skills. Additionally, the QCA publication 'Introduction to Key Skills' will be helpful.

Other subjects may be used to develop your key skills as well as AS and A2 Biology.

Remember that key skills are in demand by Further Education institutions and by employers.

What skills will I need?

For AS Biology, you will be tested by assessment objectives: these are the skills and abilities that you should have acquired by studying the course. The assessment objectives for AS Biology are shown below.

Knowledge with understanding

- recall of facts, terminology and relationships
- understanding of principles and concepts
- drawing on existing knowledge to show understanding of the responsible use of biological applications in society
- selecting, organising and presenting information clearly and logically

Application of knowledge and understanding, analysis and evaluation

- explaining and interpreting principles and concepts
- interpreting and translating, from one to another, data presented as continuous prose or in tables, diagrams and graphs
- carrying out relevant calculations
- applying knowledge and understanding to familiar and unfamiliar situations
- assessing the validity of biological information, experiments, inferences and statements

You must also present arguments and ideas clearly and logically, using specialist vocabulary where appropriate. Remember to balance your argument!

Experimental and investigative skills

Biology is a practical subject and part of the assessment of AS Biology will test your practical skills. This may be done during your lessons or may be tested in a more formal practical examination. You will be tested on four main skills:

- planning
- implementing
- analysing evidence and drawing conclusions
- evaluating evidence and procedures.

The skills may be assessed in the context of separate practical exercises, although more than one skill may be assessed in any one exercise. They may also be assessed all together in the context of a 'whole investigation'. An investigation may be set by your teacher or you may be able to pursue an investigation of your own choice.

You will receive guidance about how your practical skills will be assessed from your teacher. This study guide concentrates on preparing you for the written examinations testing the subject content of AS Biology.

Different types of questions in AS examinations

In AS Biology examinations different types of questions are used to assess your abilities and skills. Unit tests mainly use structured questions requiring both short-answers and more extended answers.

Short-answer questions

A question will normally begin with a brief amount of stimulus material. This may be in the form of a diagram, data or graph. A short-answer question may begin by testing recall. Usually this is followed up by questions which test understanding. Often you will be required to analyse data.

Short-answer questions normally have a space for your responses on the printed paper. The number of lines is a guide as to the amount of words you will need to answer the question. The number of marks indicated on the right side of the paper shows the number of marks you can score for each question part.

Here are some examples. (The answers are shown in blue)

The diagram shows part of a DNA molecule.

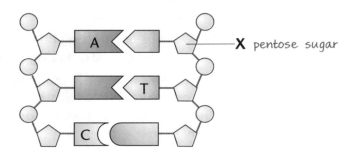

X pentose sugar

(a) Label part X.   [1]

(b) Complete the diagram by writing a letter for each missing organic base in each empty box.   [1]

(b) How do two strands of DNA join to each other?

The organic bases ✓ link the strands by hydrogen bonds ✓   [2]

Structured questions

Structured questions are in several parts. The parts are usually about a common context and they often progress in difficulty as you work through each of the parts. They may start with simple recall, then test understanding of a familiar or unfamiliar situation. If the context seems unfamiliar the material will still be centred around concepts and skills from the Biology specification. (If a student can answer questions about unfamiliar situations then they display understanding rather than simple recall.)

The most difficult part of a structured question is usually at the end. Ascending in difficulty, a question allows a candidate to build in confidence. Right at the end technological and social applications of biological principles give a more demanding challenge. Most of the questions in this book are structured questions. This is the main type of question used in the assessment of AS Biology.

When answering structured questions, do not feel that you have to complete a question before starting the next. Answering a part that you are sure of will build your confidence. If you run out of ideas go on to the next question. This will be more profitable than staying with a very difficult question which slows down progress. Return at the end when you have more time.

Here is an example of a structured question which becomes progressively more demanding.

Question

The diagram shows the molecules of a cell surface membrane.

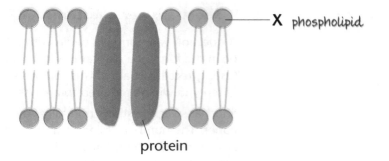

X phospholipid

protein

(a) (i)   Label molecule X.   [1]

    (ii) The part of molecule X facing the outside of a cell is hydrophilic. What does this mean?

       water loving/water attracting   [1]

    (iii) Describe **one** feature of the part of molecule X which faces inwards.

       Hydrophobic/ water hating fatty acid residues   [1]

(b) Explain how the protein shown in the diagram can actively transport the glucose molecule into the cell.

    Energy is released from mitochondria near the channel protein the channel protein opens   [3]

Note the help given in diagrams. The labelling of the protein molecule may trigger the memory so that the candidate has to make a small step to link the 'channel' function to this diagram. Examiners give clues! Expect more clues at AS Level than at A2 Level.

Extended answers

In AS Biology, questions requiring more extended answers will usually form part of structured questions. They will normally appear at the end of a structured question and will typically have a value of three to six marks. Longer answers are allocated more lines, so you can use this as a guide as to the extent of your answer. The mark allocation is a guide as to how many points you need to make in your response. Often for an answer worth six marks the mark scheme could have eight creditable answers. You are awarded up to the maximum, six in this instance.

Depending on the awarding body, longer extended questions may be set. These are often open response questions. These questions may be worth up to ten marks for full credit. Extended answers are used to allocate marks for the quality of written communication.

Candidates are assessed on their ability to use a suitable style of writing, and organise relevant material, both logically and clearly. The use of specialist biological words in context is also assessed. Spelling, punctuation and grammar are also taken into consideration. Here is a longer extended response question.

Question

Give an account of the effects of sewage entry into a river and explain the possible consequences to organisms downstream.

The sewage enters the river and is decomposed by bacteria. ✔ *These bacteria are saprobiotic* ✔ *they produce nitrates which act as a fertiliser.* ✔ *Algae form a blanket on the surface* ✔ *light cannot reach plants under the algae so these plants die.* ✔ *Bacteria decompose the dead plants* ✔ *the bacteria use oxygen/bacteria are aerobic* ✔ *fish die due to lack of oxygen* ✔ *Tubifex worms or bloodworms increase near sewage entry* ✔ *mayfly larvae cannot live close to sewage entry/mayfly larvae appear a distance downstream where oxygen levels return.* ✔

10 marking points → [7]

Remember that mark schemes for extended questions often exceed the question total, but you can only be awarded credit up to a maximum. Examiners sometimes build in a hurdle, e.g. in the above responses, references to one organism which increases in population is worthy of a mark, and another which decreases in population is worth another. Continually referring to different species which repeat a growth pattern will not gain further credit.

Exam technique

AS Biology builds from grade C in GCSE Science Double Award or the equivalent in Science: Biology. This study guide has been written so that you will be able to tackle AS Biology from a GCSE Science background.

You should not need to search for important Biology from GCSE science because this has been included where needed in each chapter. If you have not studied Science for some time, you should still be able to learn AS Biology using this text alone.

What are examiners looking for?

Whatever type of question you are answering, it is important to respond in a suitable way. Examiners use instructions to help you to decide the length and depth of your answer. The most common words used are given below, together with a brief description of what each word is asking for.

Define

This requires a formal statement. Some definitions are easy to recall.

Define the term active transport.

This is the movement of molecules from where they are in lower concentration to where they are in higher concentration. The process requires energy.

Other definitions are more complex. Where you have problems it is helpful to give an example.

Define the term endemic.

This means that a disease is found regularly in a group of people, district or country. Use of an example clarifies the meaning. Indicating that malaria is invariably found everywhere in a country, confirms understanding.

Explain

This requires a reason. The amount of detail needed is shown by the number of marks allocated.

Explain the difference between resolution and magnification.

Resolution is the ability to be able to distinguish between two points whereas magnification is the number of times an image is bigger than an object itself.

State

This requires a brief answer without any reason.

State one role of blood plasma in a mammal.

Transport of hormones to their target organs.

List

This requires a sequence of points with no explanation.

List the abiotic factors which can affect the rate of photosynthesis in pond weed.

carbon dioxide concentration; amount of light; temperature; pH of water

Describe

This requires a piece of prose which gives key points. Diagrams should be used where possible.

Describe the nervous control of heart rate.

The medulla oblongata ✓ of the brain connects to the sino atrial node in the right atrium, wall ✓ via the vagus nerve and the sympathetic nerve ✓ the sympathetic nerve speeds up the rate ✓ the vagus nerve slows it down. ✓

Discuss

This requires points both for and against, together with a criticism of each point. (**Compare** is a similar command word.)

Discuss the advantages and disadvantages of using systemic insecticides in agriculture.

Advantages are that the insecticides kill the pests which reduce yield ✓ they enter the sap of the plants so insects which consume sap die ✓ the insecticide lasts longer than a contact insecticide, 2 weeks is not uncommon ✓

Disadvantages are that insecticide may remain in the product and harm a consumer e.g. humans ✓ it may destroy organisms other than the target ✓ no insecticide is 100% effective and develops resistant pests. ✓

Suggest

This means that there is no single correct answer. Often you are given an unfamiliar situation to analyse. The examiners hope for logical deductions from the data given and that, usually, you apply your knowledge of biological concepts and principles.

The graph shows that the population of lynx decreased in 1980. Suggest reasons for this.

Weather conditions prevented plant growth ✓ so the snowshoe hares could not get enough food and their population remained low ✓ so the lynx did not have enough hares(prey) to predate upon. ✓ The lynx could have had a disease which reduced numbers. ✓

Calculate

This requires that you work out a numerical answer. Remember to give the units and to show your working, marks are usually available for a partially correct answer. If you work everything out in stages write down the sequence. Otherwise if you merely give the answer and it is wrong, then the working marks are not available to you.

Calculate the Rf value of spot X. (X is 25 mm from start and solvent front is 100 mm)

$$Rf = \frac{\text{distance moved by spot}}{\text{distance moved by the solvent front}}$$

$$= \frac{25 \text{ mm}}{100 \text{ mm}}$$

$$= 0.25$$

Outline

This requires that you give only the main points. The marks allocated will guide you on the number of points which you need to make.

Outline the use of restriction endonuclease in genetic engineering.

The enzyme is used to cut the DNA of the donor cell. ✔

It cuts the DNA up like this A T $\vert$ G C C G A T = A T   + G C C G A T ✔
                      T A C G G C $\vert$ T A   T A C G G C   T A

The DNA in a bacterial plasmid is cut with the same restriction endonuclease. ✔

The donor DNA will fit onto the sticky ends of the broken plasmid. ✔

If a question does not seem to make sense, you may have mis-read it. Read it again!

Some dos and don'ts

Dos

Do *answer the question*

No credit can be given for good Biology that is irrelevant to the question.

Do *use the mark allocation to guide how much you write*

Two marks are awarded for two valid points – writing more will rarely gain more credit and could mean wasted time or even contradicting earlier valid points.

Do *use diagrams, equations and tables in your responses*

Even in 'essay style' questions, these offer an excellent way of communicating biology.

Do *write legibly*

An examiner cannot give marks if the answer cannot be read.

Do *write using correct spelling and grammar. Structure longer essays carefully*

Marks are now awarded for the quality of your language in exams.

Don'ts

Don't *fill up any blank space on a paper*

In structured questions, the number of dotted lines should guide the length of your answer.

If you write too much, you waste time and may not finish the exam paper. You also risk contradicting yourself.

Don't *write out the question again*

This wastes time. The marks are for the answer!

Don't *contradict yourself*

The examiner cannot be expected to choose which answer is intended. You could lose a hard-earned mark.

Don't *spend too much time on a part that you find difficult*

You may not have enough time to complete the exam. You can always return to a difficult calculation if you have time at the end of the exam.

What grade do you want?

Everyone would like to improve their grades but you will only manage this with a lot of hard work and determination. You should have a fair idea of your natural ability and likely grade in biology and the hints below offer advice on improving that grade.

For a Grade A

You will need to be a very good all-rounder.

- You must go into every exam knowing the work extremely well.
- You must be able to apply your knowledge to new, unfamiliar situations.
- You need to have practised many, many exam questions so that you are ready for the type of question that will appear.

The exams test all areas of the syllabus and any weaknesses in your biology will be found out. There must be no holes in your knowledge and understanding. For a Grade A, you must be competent in all areas.

For a Grade C

You must have a reasonable grasp of biology but you may have weaknesses in several areas and you will be unsure of some of the reasons for the biology.

- Many Grade C candidates are just as good at answering questions as the Grade A students but holes and weaknesses often show up in just some topics.
- To improve, you will need to master your weaknesses and you must prepare thoroughly for the exam. You must become a better all-rounder.

For a Grade E

You cannot afford to miss the easy marks. Even if you find biology difficult to understand and would be happy with a Grade E, there are plenty of questions in which you can gain marks.

- You must memorise all definitions.
- You must practise exam questions to give yourself confidence that you do know some biology. In exams, answer the parts of questions that you know first. You must not waste time on the difficult parts. You can always go back to these later.
- The areas of biology that you find most difficult are going to be hard to score on in exams. Even in the difficult questions, there are still marks to be gained. Show your working in calculations because credit is given for a sound method. You can always gain some marks if you get part of the way towards the solution.

What marks do you need?

The table below shows how your average mark is transferred into a grade.

average	80%	70%	60%	50%	40%
grade	A	B	C	D	E

Four steps to successful revision

Step 1: Understand

- Study the topic to be learned slowly. Make sure you understand the logic or important concepts.
- Mark up the text if necessary – underline, highlight and make notes.
- Re-read each paragraph slowly.

GO TO STEP 2

Step 2: Summarise

- Now make your own revision note summary:
  What is the main idea, theme or concept to be learned?
  What are the main points? How does the logic develop?
  Ask questions: Why? How? What next?
- Use bullet points, mind maps, patterned notes.
- Link ideas with mnemonics, mind maps, crazy stories.
- Note the title and date of the revision notes
  (e.g. Biology: Cells, 3rd March).
- Organise your notes carefully and keep them in a file.

This is now in **short term memory**. You will forget 80% of it if you do not go to Step 3.
GO TO STEP 3, but first take a 10 minute break.

Step 3: Memorise

- Take 25 minute learning 'bites' with 5 minute breaks.
- After each 5 minute break test yourself:
  Cover the original revision note summary.
  Write down the main points.
  Speak out loud (record on tape).
  Tell someone else.
  Repeat many times.

The material is well on its way to **long term memory**.
You will forget 40% if you do not do step 4. GO TO STEP 4

Step 4: Track/Review

- Create a Revision Diary (one A4 page per day).
- Make a revision plan for the topic, e.g. 1 day later, 1 week later, 1 month later.
- Record your revision in your Revision Diary, e.g.
  Biology: Cells, 3rd March 25 minutes
  Biology: Cells, 5th March 15 minutes
  Biology: Cells, 3rd April 15 minutes
  ... and then at monthly intervals.

Chapter 1
Biological molecules

The following topics are covered in this chapter:

- Essential substances
- Carbohydrates
- Lipids
- Proteins
- Biochemical tests and chromatography
- The importance of water to life

1.1 Essential substances

After studying this section you should be able to:

- recall the number of elements essential for life
- recall the four major elements and how they are linked to form biological molecules

LEARNING SUMMARY

Substances required for living processes

AQA A	M1
AQA B	M1
EDEXCEL	M1
OCR	M1
WJEC	M1
NICCEA	M1

Element	percentage (approximate)
Carbon	9.5
Hydrogen	63.0
Oxygen	25.5
Nitrogen	1.4
Calcium	0.32
Potassium	0.06
Phosphorus	0.20
Chlorine	0.03
Sulphur	0.05
Sodium	0.03

Living things are based on a total of 16 elements out of the 92 which exist on Earth. Over 99% of the biomass of organisms is composed of just 4 key elements, carbon, hydrogen, oxygen, and nitrogen.

Carbon is the most important element because of its following properties:

- carbon atoms bond with each other in long chains
- the chains can be branched or even joined up as rings
- the carbon atoms bond with other important elements like hydrogen, oxygen, nitrogen, sulphur, calcium and phosphorus.

The linking of carbon to carbon in long chains forms the backbone of important structural molecules. Electrons not used in the bonding of carbon to carbon are shared with other elements, like hydrogen, oxygen and nitrogen. All the essential elements together have incredible properties contributing to the diversity of life forms on Earth.

The table in the margin shows a range of elements found in the human body.

Some important elements

The molecules of every organism consist of a number of elements which bond together, and are vital to life. The properties of these inter-linking elements contribute to both structure and life processes.

Sodium atoms and chlorine atoms would be very destructive to life! Their properties would kill cells. However, in the form of ions they are vital to the survival of an organism. Inorganic ions are needed by organisms for a range of functions.

Ion	example in animals	examples in plants
Iron (Fe^{2+})	haemoglobin – transports oxygen efficiently; cytochromes – energy release	chlorophyll synthesis; energy release
Magnesium (Mg^{2+})	muscle and nerve function; bone formation	major constituent of chlorophyll
Potassium (K^+)	transmission of nerve impulses and muscle function; formation and disease resistance	chlorophyll formation
Calcium (Ca^{2+})	formation of bones and teeth; clotting of blood; muscle function	cell wall constituent; cell division

1.2 Carbohydrates

After studying this section you should be able to:

- recall the main elements found in carbohydrates
- recall the structure of glucose, starch, galactose and maltose
- recall the role of glucose, starch, cellulose and pectin

Structure of carbohydrates

AQA A	M1
AQA B	M1
EDEXCEL	M1
OCR	M1
WJEC	M1
NICCEA	M1

Monosaccharides

All carbohydrates are formed from the elements carbon (C), hydrogen (H) and oxygen (O). The formula of a carbohydrate is always $(CH_2O)_n$. The n represents the number of times the basic CH_2O unit is repeated, e.g. where n = 6 the molecular formula is $C_6H_{12}O_6$. This is the formula shared by glucose and other simple sugars like fructose. These simple sugars are known as monosaccharides.

The molecular formula, $C_6H_{12}O_6$, does not indicate how the atoms bond together. Bonded to the carbon atoms are a number of $-H$ and $-OH$ groups. Different positions of these groups on the carbon chain are responsible for different properties of the molecules. The structural formulae of α and β glucose are shown below.

These molecules are mirror images of each other. When molecules have the same molecular formula but different structural formulae, they are known as **isomers**. Isomers have different properties to each other.

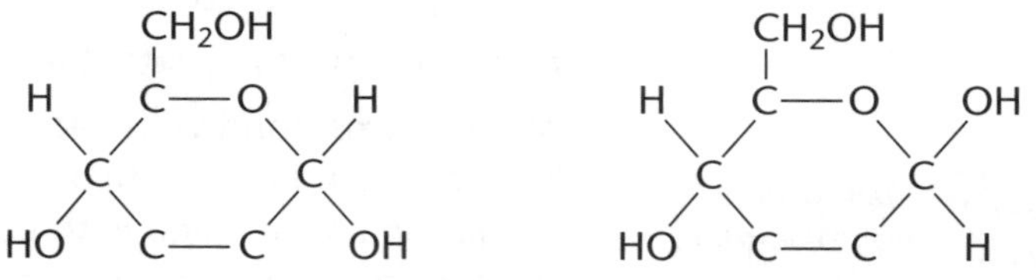

α glucose   β glucose

Glucose is so small that it can pass through the villi and capillaries into our bloodstream. The molecules subsequently release energy as a result of respiration. Simple glucose molecules are capable of so much more. They can combine with others to form bigger molecules.

Disaccharides

Each glucose unit is known as a monomer and is capable of linking others. This diagram shows two molecules of α glucose forming a disaccharide.

In your examinations look for different monosaccharides being given, like fructose or β glucose. You may be asked to show how they bond together. The principle will be exactly the same.

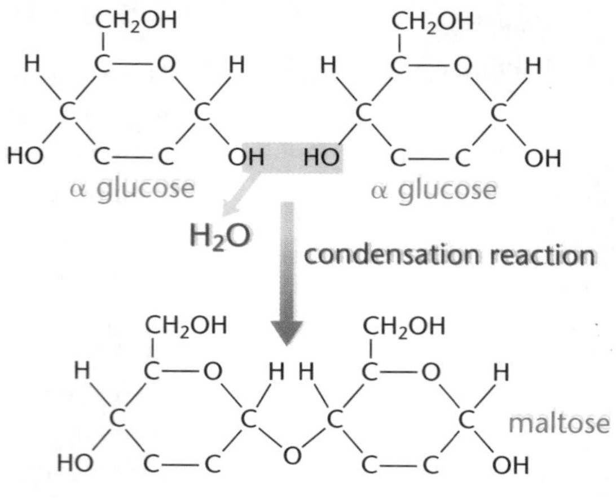

α glucose   α glucose
H_2O condensation reaction
maltose

A condensation reaction means that as two carbohydrate molecules bond together a water molecule is produced. The link formed between the two glucose molecules is known as a glycosidic bond.

A glycosidic bond can also be broken down to release separate monomer units. This is the opposite of the reaction shown above. Instead of water being given off, a water molecule is needed to break each glycosidic bond. This is called hydrolysis because water is needed to split up the bigger molecule.

'Lysis' literally means 'splitting'. In hydrolysis water is needed in the reaction to break down the molecule.

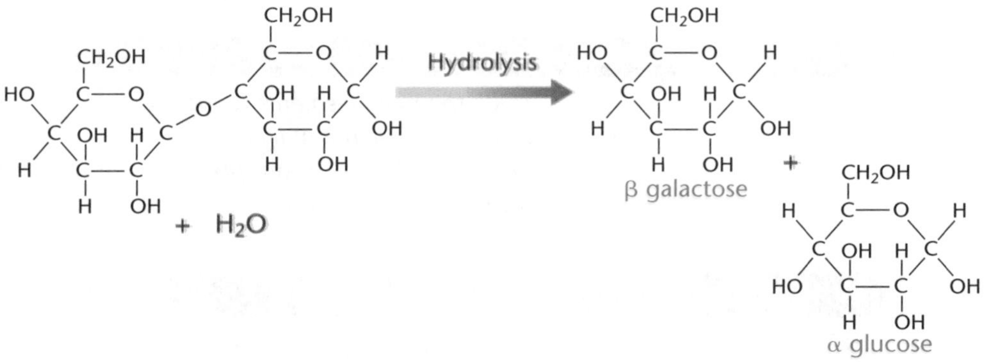

β galactose

α glucose

Polysaccharides

Like disaccharides, they consist of monomer units linked by the glycosidic bond. However, instead of just two monomer units they can have many. Chains of these 'sugar' units are known as polymers. These larger molecules have important structural and storage roles.

Starch is a polymer of the sugar, glucose. The diagram below shows part of a starch molecule.

Notice the five glycosidic bonds on just a small part of a starch molecule.

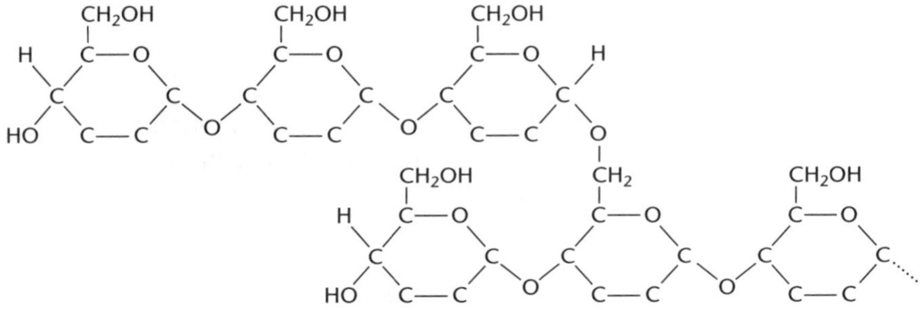

part of a branched section of a starch molecule

The table classifies carbohydrates.

Remember that each individual sugar unit is a monomer.

Monosaccharide (one sugar unit)	Disaccharide (two sugar units)	Polysaccharide (many sugar units)
glucose	maltose	starch
fructose	sucrose	glycogen
galactose	lactose	cellulose
		pectins

How useful are polysaccharides?

- Starch is stored in organisms as a future energy source, e.g. potato has a high starch content to supply energy for the buds to grow at a later stage.
- Glycogen is stored in the liver, which releases glucose for energy in times of low blood sugar.

Both starch and glycogen are insoluble which enables them to remain inside cells.

- Cellulose has long chains and branches which help form a tough protective layer around plant cells, the cell wall.
- Pectins are used alongside cellulose in the cell wall. They are polysaccharides which are bound together by calcium pectate.

Pectins help cells to bind together.

Together the cellulose and pectins give exceptional mechanical strength. The cell wall is also permeable to a wide range of substances.

23

1.3 Lipids

After studying this section you should be able to:

- *recall the main elements found in lipids*
- *recall the structure of lipids*
- *distinguish between saturated and unsaturated fats*
- *recall the role of lipids*

LEARNING SUMMARY

What are lipids?

AQA A   M1
AQA B   M1
EDEXCEL   M1
OCR   M1
WJEC   M1
NICCEA   M1

These are the **oils**, **fats** and **waxes**. They consist of exactly the same elements as carbohydrates, i.e. carbon (C), hydrogen (H) and oxygen (O) but their proportion is different. There is always a high proportion of carbon and hydrogen, with a small proportion of oxygen. The diagram below shows the structural formula of a typical fat.

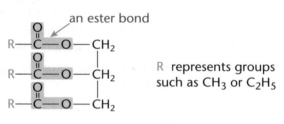

a **triglyceride** fat

Fats and oils are formed when fatty acids react with glycerol. During this reaction water is produced, a further example of a condensation reaction. The essential bond is the **ester bond**.

Note that water is produced during triglyceride formation. This is another example of a condensation reaction. Different triglyceride fats are formed from different fatty acids.

3 fatty acids   glycerol   a triglyceride fat   water

Fats and oils can be changed back into the original fatty acids and glycerol. Enzymes (see page 46) are needed for this transformation together with water molecules. An enzyme reaction which requires water to break up a molecule is known as **hydrolysis**.

What are saturated and unsaturated fats?

The answer lies in the types of fatty acid used to produce them.

The hydrocarbon chains are so long that they are often represented by the acid group (–COOH) and a zig-zag line.

    unsaturated
〜〜〜=〜〜〜COOH

    saturated
〜〜〜〜〜〜COOH

stearic acid

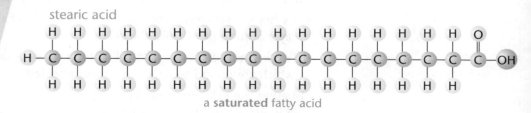

a **saturated** fatty acid

oleic acid

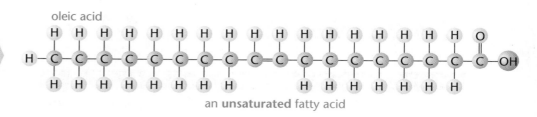

an **unsaturated** fatty acid

> **KEY POINT**
> Saturated fatty acids have no C=C (double bonds) in their hydrocarbon chain, but unsaturated fatty acids do. This is the difference.

How useful are lipids?

Like carbohydrates, they are used as an energy supply, but a given amount of lipid release more energy than the same amount of carbohydrate. Due to their insolubility in water and compact structure, lipids have long-term storage qualities. Adipose cells beneath our skin contain large quantities of fat which insulate us and help to maintain body temperature. Fat gives mechanical support around our soft organs and even gives electrical insulation around our nerve axons.

An aquatic organism such as a dolphin has a large fat layer which:

- is an energy store
- a thermal insulator
- helps the animal remain buoyant.

The most important role of lipids is their function in cell membranes. To fulfil these functions a triglyceride fat is first converted into a phospholipid.

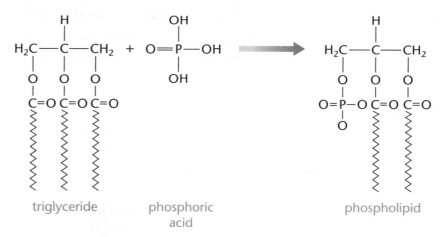

triglyceride   phosphoric acid   phospholipid

Phosphoric acid replaces one of the fatty acids of the triglyceride. The new molecule, the phospholipid, is a major component of cell membranes. The diagram below represents a phospholipid.

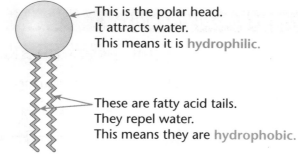

This is the polar head. It attracts water. This means it is hydrophilic.

These are fatty acid tails. They repel water. This means they are hydrophobic.

a phospholipid

1.4 Proteins

After studying this section you should be able to:

- recall the main elements found in proteins
- recall how proteins are constructed
- recall the structure of proteins
- recall the major functions of proteins

LEARNING SUMMARY

The building blocks of proteins

AQA A	M1
AQA B	M1
EDEXCEL	M1
OCR	M1
WJEC	M1
NICCEA	M1

Just like the earlier carbohydrate and lipid molecules 'R' represents groups such as –CH$_3$ and –C$_2$H$_5$. There are about 20 commonly found amino acids but you will not need to know them all. Instead, learn the basic structure shown opposite.

Like carbohydrates and lipids, proteins contain the elements carbon (C), hydrogen (H) and oxygen (O), but in addition they also **always** contain **nitrogen** (N).

Sulphur is often present as well as iron and phosphorus. Before understanding how proteins are constructed, the structure of **amino acids** should be noted. The diagram below shows the general structure of an amino acid.

an amino acid

How is a protein constructed?

This is another example of a condensation reaction as water is produced as the dipeptide molecule is assembled.

Note that the peptide bonds can be broken down by a hydrolysis reaction.

The process begins by amino acids bonding together. The diagram shows two amino acids being joined together by a **peptide bond**.

amino acid   *amino acid*   *a dipeptide* + H$_2$O

The sequence of amino acids along a polypeptide is controlled by another complex molecule, DNA (see the genetic code, page 82).

When many amino acids join together a long-chain **polypeptide** is produced. The linking of amino acids in this way takes place during protein synthesis (see page 83). There are around 20 different amino acids. Organisms join amino acids in different linear sequences to form a variety of polypeptides, then build these polypeptides into complex molecules, the **proteins**. Humans need eight essential amino acids as adults and ten as children, all the others can be made inside the cells.

The structure of proteins

AQA A	M1
AQA B	M1
EDEXCEL	M1
OCR	M1
WJEC	M1
NICCEA	M1

Primary protein structure

This is the **linear sequence** of amino acids.

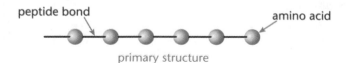

primary structure

Secondary protein structure

Polypeptides become twisted or coiled. These shapes are known as the secondary structure. There are two common secondary structures; the α-helix and the β-pleated sheet.

Both secondary structures give additional strength to proteins. The α-helix helps make tough fibres like the protein in your nails, e.g. keratin. The β-pleated sheet helps make the strength-giving protein in silk, fibroin. Many proteins are made from both α-helix and β-pleated sheet.

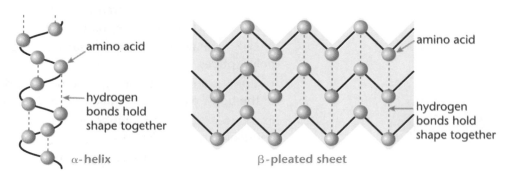

amino acid
hydrogen bonds hold shape together
α-helix

amino acid
hydrogen bonds hold shape together
β-pleated sheet

The polypeptides are held in position by hydrogen bonds. In both α-helices and β-pleated sheets the C=O of one amino acid bonds to the H–N of an adjacent amino acid, like this, C=O --- H–N.

hydrogen bonds

An α-helix is a tight, twisted strand; a β-pleated sheet is where a zig-zag line of amino acids bonds with the next, and so on. This forms a sheet or ribbon shape.

The protein shown, only achieves a secondary structure as the simple α–helix polypeptides do not undergo further folding.

This is the structure of a fibrous protein. It is made of three α–helix polypeptides twisted together.

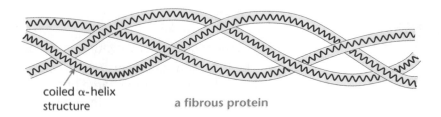

coiled α-helix structure

a fibrous protein

Tertiary protein structure

Note that the specific contours of proteins have extremely significant roles in life processes. (See enzymes page 46 and immunity page 134.)

This is when a polypeptide is folded into a precise shape. The polypeptide is held in 'bends' and 'tucks' in a permanent shape by a range of bonds including:

* disulphide bridges (sulphur–sulphur bonds)
* hydrogen bonds
* ionic bonds.

This is the structure of a **globular** protein. It is made of an α-helix and a β-pleated sheet. Precise shapes are formed with specific contours.

Quaternary protein structure

Some proteins consist of different polypeptides bonded together to form extremely intricate shapes. A haemoglobin molecule is formed from four separate polypeptide chains. It also has a haem group, which contains iron. This inorganic group is known as a prosthetic group and in this instance aids oxygen transport.

Note that some proteins do not have a quaternary structure. If they consist of just one folded polypeptide then they are classified as having tertiary structure. If they are simple fibres of α-helices or β-pleated sheets then they have only secondary protein structure.

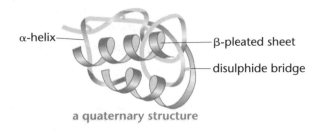

α-helix
β-pleated sheet
disulphide bridge

a quaternary structure

How useful are proteins?

AQA A   M1
AQA B   M1
EDEXCEL   M1
OCR   M1
WJEC   M1
NICCEA   M1

Just as carbohydrates and lipids can release energy, proteins are just as beneficial. Broken down into their component amino acids, these also liberate energy during respiration. The list below shows important uses of proteins:

- cell-membrane proteins transport substances across the membrane for processes such as facilitated diffusion and active transport
- enzymes catalyse biochemical reactions, e.g. pepsin breaks down protein into polypeptides
- hormones are passed through the blood and trigger reactions in other parts of the body, e.g. insulin regulates blood sugar
- immuno-proteins, e.g. antibodies are made by lymphocytes and act against antigenic sites on microbes
- structural proteins give strength to organs, e.g. collagen makes tendons tough
- transport proteins, e.g. haemoglobin transports oxygen in the blood
- contractile proteins, e.g. actin and myosin help muscles shorten during contraction
- storage proteins, e.g. aleurone in seeds helps germination, and casein in milk helps supply valuable protein to babies
- buffer proteins, e.g. blood proteins, due to their charge, help maintain the pH of plasma.

Progress check

1 List the sequence of structures in a globular protein such as haemoglobin.

2 The following statements refer to proteins used for different functions in the body. The list gives the name of different types of protein. Link the name of each type of protein with the correct statement.

(i) transport proteins   (vi) contractile proteins
(ii) immuno-proteins   (vii) enzymes
(iii) storage proteins   (viii) structural proteins
(iv) buffer proteins   (ix) hormones
(v) cell-membrane proteins

A used to transport substances across the membrane for processes such as facilitated diffusion.

B used to catalyse biochemical reactions, e.g. amylase breaks down starch into maltose.

C passed through blood, used to trigger reactions in other parts of the body, e.g. FSH stimulates a primary follicle.

D antibodies made by lymphocytes against antigens.

E used to give strength to organs, e.g. collagen makes tendons tough.

F haemoglobin is used to transport oxygen in blood.

G actin and myosin help muscles shorten during contraction.

H aleurone in seeds is a source of amino acids as it is broken down during germination.

I blood proteins, due to their charge, help maintain the pH of plasma.

1 primary structure: amino acids linked in a linear sequence; secondary structure: α-helix or β-pleated sheet; tertiary structure: further folding of polypeptide held by disulphide bridges, ionic bonds, and hydrogen bonds; quaternary structure: two or more polypeptides bonded together.
2 A (v), B (vii), C (ix), D (ii), E (viii), F (i), G (vi), H (iii), I (iv).

1.5 Biochemical tests and chromatography

LEARNING SUMMARY

After studying this section you should be able to:

- describe biochemical tests for carbohydrates, proteins and lipids
- describe the separation and identification of molecules by chromatography

Biochemical tests

AQA A	M1
AQA B	M1
EDEXCEL	M1
OCR	M1
WJEC	M1
NICCEA	M1

> All the biochemical tests need to be learned. This work is good value because they are regularly tested in 2 or 3 mark question components.

Tests for carbohydrates in the laboratory

Benedict's test used to identify reducing sugars (monosaccharides and some disaccharides)

- Add Benedict's solution to the chemical sample and heat.
- The solution changes from blue to brick-red or yellow if a reducing sugar is present.

Non-reducing sugar test used to test for non-reducing sugars, e.g. the disaccharide, sucrose

- First a Benedict's test is performed.
- If the Benedict's test is negative, the sample is hydrolysed by heating with hydrochloric acid, then neutralised with sodium hydrogen carbonate.
- This breaks the glycosidic bond of the disaccharide, releasing the monomers.
- A second Benedict's test is performed which will be positive because the monomers are now free.

Starch test

- Add iodine solution to the sample.
- If starch is present the colour changes to blue-black.

Tests for lipids in the laboratory

Emulsion test used to identify fats and oils

- Add ethanol to the sample, shake, then pour the mixture into water.
- If fats or oils are present then a white emulsion appears at the surface.

Tests for proteins in the laboratory

Biuret test used to identify any protein

- Add dilute sodium hydroxide and dilute copper sulphate to the sample.
- A violet colour appears if a protein is present.

Chromatography

AQA A	M1
AQA B	M1
NICCEA	M1

> You need to remember that this technique separates substances in terms of the relative size of the molecules.

This technique is used to separate out the components in a mixture. It is used to separate out the components of substances such as chlorophyll, and can be used to help identify substances. The method is outlined below:

- a spot of the substance is placed on chromatography paper and left to dry
- the paper is suspended in a solvent such as propanone
- as the solvent molecules move through the paper the components begin to move up the paper, big molecules move slower than small ones
- the small solvent molecules move through the paper faster than any of the components of the substance
- the substance separates out into different spots or bands.

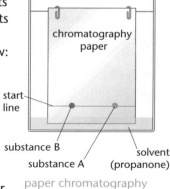

paper chromatography
(before separation)

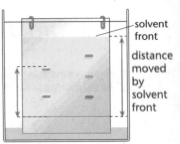

solvent front

distance moved by solvent front

substance B had two component compounds

substance A had three component compounds

paper chromatography (after separation)

R$_f$ value of a substance

This is calculated after the distances moved by compounds and solvent up the chromatogram have been measured. The distance moved by the solvent is called the solvent front.

$$R_f \text{ value} = \frac{\text{distance moved by substance}}{\text{distance moved by solvent front}}$$

Different compounds show up as different coloured bands. The technique shows the number of compounds in the mixture. When this method is used on chlorophyll, five colours separate out.

> The longer the chromatogram is left after the start, the higher the spots or bands ascend. For this reason every compound has its R$_f$ value calculated. However long the chromatogram is left, the R$_f$ value is the same when using the same solvent.

Progress check

A chromatogram was prepared for substance X. Five different spots were noted on the chromatogram.

1 What does this indicate?

2 What is the equation used to calculate the R$_f$ value of a spot?

1 Substance X consisted of 5 different substances.

2 R$_f$ value = $\frac{\text{distance moved by substance}}{\text{distance moved by solvent front}}$

1.6 The importance of water to life

After studying this section you should be able to:

- *recall the properties of water*
- *recall the functions of water*

LEARNING SUMMARY

Properties and uses of water

AQA A	M1
AQA B	M1
EDEXCEL	M1
OCR	M1
WJEC	M1
NICCEA	M1

> Try to learn all of the functions of water molecules given in the list. Water is used in so many ways that the chance of being questioned on the topic is high.

Water is essential to living organisms. The list below shows some of its properties and uses.

- **Hydrogen bonds** are formed between the oxygen of one water molecule and the hydrogen of another. As a result of this water molecules have an attraction for each other known as **cohesion**.

- **Cohesion** is responsible for surface tension which enables aquatic insects like pond skaters to walk on a pond surface. It also aids capillarity, the way in which water moves through xylem in plants.

- Water is a **dipolar** molecule, which means that the oxygen has a slight negative charge at one end of the molecule, and each hydrogen a slight positive charge at the other end.

- Other **polar** molecules dissolve in water. The different charges on these molecules enable them to fit into water's hydrogen bond structure. Ions in solution can be transported or can take part in reactions. Polar substances which dissolve are **hydrophilic** and non-polar, which cannot dissolve in water, **hydrophobic**.

- Water is used in **photosynthesis**, so it is responsible for the production of glucose. This in turn is used in the synthesis of many chemicals.

- Water helps in the **temperature regulation** of many organisms. It enables the cooling down of some organisms. Owing to a high **latent heat of vaporisation**, large amounts of body heat are needed to evaporate a small quantity of water. Organisms like humans cool down effectively but lose only a small amount of water in doing so.

- A relatively high level of heat is needed to raise the temperature of water by a small amount due to its high specific heat capacity. This enables organisms to control their body temperature more effectively.
- Water is a solvent for ionic compounds. A number of the essential elements required by organisms are obtained in ionic form, e.g.:
  (a) plants absorb nitrate ions (NO_3^-) and phosphate ions (PO_4^-) in solution
  (b) animals intake sodium ions (Na^+) and chloride ions (Cl^-).

Sample questions and model answers

1 Below are the structures of two glucose molecules.

(a) Complete the equation to show how the molecules react to form a glycosidic bond and the molecule produced.

Remember that you will be given molecule structures. These stimulate your memory which helps you work out the answer.

(b) Which form of glucose molecules is shown?
Give a reason for your answer. [2]

α glucose, because the −OH groups on carbon atom 1 are down

The correct answer here is condensation. A regular error in questions like this is to give the wrong reaction, i.e. hydrolysis. Revise carefully then you will make the correct choice.

(c) State the type of reaction which takes place when the two molecules shown above react together. [2]

Condensation.

2 The diagram below shows a globular protein consisting of four polypeptide chains.

α-helix —

— β-pleated sheet

— disulphide

Look out for similar structures in your examinations. The proteins given may be different, but the principles remain the same.

(a) Use your own knowledge and the information given to explain how this protein shows primary, secondary, tertiary and quaternary structure. [5]

Primary structure: it is formed from chains of amino acids; it has polypeptides made of a linear sequence of amino acids.

Secondary structure: it has an α-helix, it has a β-pleated sheet.

Tertiary structure: the polypeptides are folded, the folds are held in position by disulphide bridges.

*Quaternary structure: there are four polypeptides in this protein.
Two or more are bonded together to give a quaternary structure.*

(b) Name and describe a test which would show that haemoglobin is a protein. [3]

The Biuret test.

Take a sample of haemoglobin and add water, sodium hydroxide and copper sulphate.

Most examinations include at least one biochemical test.

The colour of the mixture shows as violet or mauve if the sample is a protein.

Practice examination questions

Try all of the questions and check your answers with the mark scheme on page 139.

1 The chromatogram below shows a substance which has been separated into its component compounds.

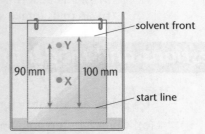

(a) Calculate the R_f value of spot Y. [2]

(b) Which spot contains the biggest molecules? [1]

(c) The chromatogram had been left for six hours after a drop was put on the start line. Why was it important to take the measurement for the calculation of the R_f value of Y before another hour had past? [1]

2 (a) Complete the equation below to show the breakdown of a triglyceride fat into fatty acids and glycerol. [2]

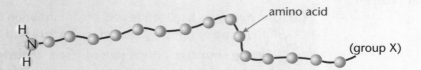

a triglyceride fat   3 molecules
of water

(b) Describe a biochemical test which would show that a sample was a fat. [3]

3 The diagram below shows a polypeptide consisting of 15 amino acids.

(a) Name the bond between each pair of amino acids in this polypeptide. [1]

(b) What is group X? [1]

(c) Which level of protein structure is shown by this polypeptide? Give a reason for your answer. [2]

4 Explain how the following properties of water are useful to living organisms:

(a) a large latent heat of evaporation [2]

(b) a high specific heat capacity [2]

(c) the cohesive attraction of water molecules for each other. [2]

Chapter 2

Cells

The following topics are covered in this chapter:

- *The ultra-structure of cells*
- *Isolation of cell organelles*
- *Specialisation of cells*

2.1 The ultra-structure of cells

After studying this section you should be able to:

- *identify cell organelles and understand their roles*
- *recall the differences between prokaryotic and eukaryotic cells*

LEARNING SUMMARY

Cell organelles

AQA A	M1
AQA B	M1
EDEXCEL	M1
OCR	M1
WJEC	M1
NICCEA	M1

The cell is the basic functioning unit of organisms in which chemical reactions take place. These reactions involve energy release needed to support life and build structures. Organisms consist of one or more cells. The amoeba is composed of one cell, whereas millions of cells make up a human.

> Every cell possesses internal coded instructions to control cell activities and development (see the genetic code page 82). Cells also have the ability to continue life by some form of cell division.

KEY POINT

> Organelles are best seen with the aid of an electron microscope.

The ultra-structure of a cell can be seen using an electron microscope. Sub-cellular units called **organelles** become visible. Each organelle has been researched to help us understand more about the processes of life.

The animal cell and its organelles

The diagram below shows the organelles found in a typical animal cell.

> A plant cell has all of the same structures plus:
> - a cellulose cell wall
> - chloroplasts (some cells)
> - a sap vacuole with tonoplast.

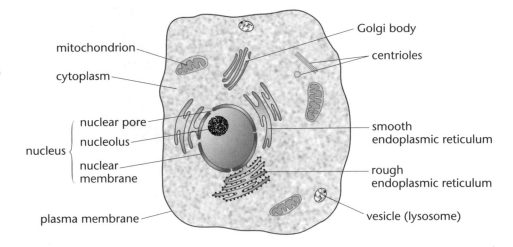

mitochondrion

cytoplasm

nuclear pore

nucleus { nucleolus

nuclear membrane

plasma membrane

Golgi body

centrioles

smooth endoplasmic reticulum

rough endoplasmic reticulum

vesicle (lysosome)

Cell surface (plasma) membrane

This covers the outside of a cell and consists of a **double** layered sheet of lipid molecules interspersed with proteins. It separates the cell from the outside environment, gives physical protection and allows the import and export of **selected** chemicals.

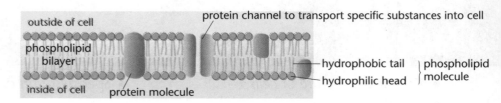

Nucleus

This controls all cellular activity using coded instructions located in DNA. These coded instructions enable the cell to make specific proteins (see protein synthesis page 83). RNA is produced in the nucleus and leaves via the nuclear pores. The nucleus stores, replicates and decodes DNA.

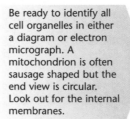

Be ready to identify all cell organelles in either a diagram or electron micrograph. A mitochondrion is often sausage shaped but the end view is circular. Look out for the internal membranes.

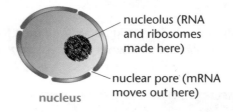

nucleolus (RNA and ribosomes made here)

nuclear pore (mRNA moves out here)

nucleus

cristae

mitochondrion

Mitochondria

These consist of an outer membrane enclosing a semi-fluid matrix. Throughout the matrix is an internal membrane, folded into cristae. The cristae and matrix contain enzymes which enable this organelle to carry out aerobic respiration. It is the key organelle in the release of energy, making ATP available to the cell.

Mitochondria are needed for many energy requiring processes in the cell, including active transport and the movement of cilia.

Cytoplasm is often seen as grey and granular. If the image is 'clear' then you are probably looking at a vacuole.

Cytoplasm

Each organelle in a cell is suspended in a semi-liquid medium, the cytoplasm. Many ions are dissolved in it. It is the site of many chemical reactions.

Look for tiny dots in the cytoplasm. They will almost certainly be ribosomes. A membrane adjacent to a line of ribosomes is probably the rough endoplasmic reticulum.

Ribosomes

There are numerous ribosomes in a cell, located along rough endoplasmic reticulum. They aid the manufacture of proteins, being the site where mRNA meets tRNA so that amino acids are bonded together.

Endoplasmic reticulum (ER)

This is found as rough ER (with ribosomes) and smooth ER (without ribosomes). It is a series of folded internal membranes. Substances are transported in the spaces between the ER. The smooth ER aids the synthesis and transport of lipids.

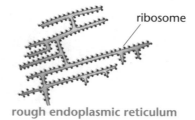

ribosome

rough endoplasmic reticulum

smooth endoplasmic reticulum

Golgi body

Look for vesicles 'pinching off' the main Golgi sacs.

This is a series of flattened sacs, each separated from the cytoplasm by a membrane. The Golgi body is a packaging system where important chemicals become membrane wrapped, forming vesicles. The vesicles become detached from the main Golgi sacs, enabling the isolation of chemicals from each other in the cytoplasm. The Golgi body aids the production and secretion of many proteins, carbohydrates and glycoproteins. Vesicle membranes merge with the plasma membrane to enable secretions to take place.

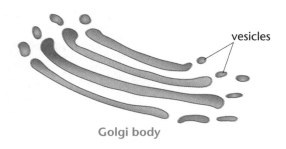

Golgi body

Lysosomes

These are specialised vesicles because they contain digestive enzymes. The enzymes have the ability to break down proteins and lipids. If the enzymes were free to react in the cytoplasm then cell destruction would result.

lysosome   centrioles

Centrioles

In a cell there are two short cylinders which contain microtubules. Their function is to aid cell division. During division they move to opposite poles as the spindle develops.

The plant cell and its organelles

All of the structures described for animal cells are also found in plant cells. Additionally there are three extra structures shown in the diagram below.

Did you spot the three extra structures in the plant cell? Remember that a root cell under the soil will not possess chloroplasts. Nor does every plant cell above the soil have chloroplasts.

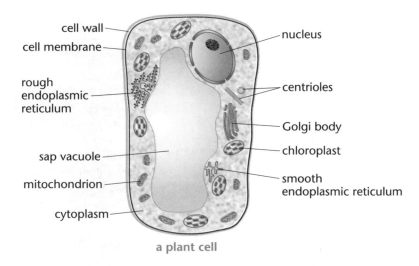

a plant cell

Cell wall

Cells other than plant cells can have cell walls, e.g. bacteria have polysaccharides other than cellulose.

Around the plasma membrane of plant cells is the cell wall. This is secreted by the cell and consists of cellulose microfibrils embedded in a layer of calcium pectate and hemicelluloses. Between the walls of neighbouring cells calcium pectate cements one cell to the next in multi-cellular plants. Plant cell walls provide a rigid support for the cell but allow many substances to be imported or exported by the cell. The wall allows the cell to build up an effective hydrostatic skeleton. When the cell has a maximum amount of water content, it has maximum strength, and is said to be turgid. Some plant cells have a cytoplasmic link which crosses the wall. These links of cytoplasm are known as plasmodesmata.

plasmodesmata

a chloroplast

Chloroplasts

These enable the plant to photosynthesise, making glucose. Each consists of an outer covering of two membranes. Inside are more membranes stacked in piles called grana. The membranes enclose a substance vital to photosynthesis, chlorophyll. Inside the chloroplast is a matrix known as the stroma which is also involved in photosynthesis.

Sap vacuole

This is a large space in a plant cell, containing chemicals such as glucose and mineral ions in water. This solution is the sap. It is surrounded by a membrane known as the tonoplast. It is important that a plant cell contains enough water to maintain internal hydrostatic pressure. When this is achieved the cell is turgid, having maximum hydrostatic strength.

Prokaryotic and eukaryotic cells

AQA A	M1
AQA B	M1
EDEXCEL	M1
OCR	M1
WJEC	M1
NICCEA	M1

Organisms can be classified into two groups, prokaryotic or eukaryotic according to their cellular structure.

> **KEY POINT**
>
> The former type of cell is characteristic of two groups of organisms, bacteria and blue-green algae. Prokaryotic cells are less complex than the eukaryotic ones and are considered to have evolved earlier.

The table below states similarities and differences between the two types of organism.

In an examination you will often be given a diagram of a cell from an organism you have not seen before. This is not a problem! The examiners are testing your recognition of the organelles found in typical prokaryotic and eukaryotic organisms.

		Prokaryotic cells	*Eukaryotic cells*
Kingdom		Prokaryotae	Protoctista, Fungi, Animalia, Plantae
Organelles	1	small ribosomes	large ribosomes
	2	DNA present but there is no nuclear membrane	DNA is enclosed in a membrane i.e. has nucleus, mitochondria, (Golgi body vesicles and ER are present)
	3	cell wall present consisting of mucopeptides	cell walls present in plant cells – cellulose cell walls present in fungi – chitin
	4	if cells have flagellae there is no 9 + 2 microtubule arrangement	if cells have flagellae there is a 9 + 2 microtubule arrangement

Progress check

1 Describe the function of each of the following cell organelles:

    nucleus   centrioles   Golgi body
    mitochondria   ribosomes   cell (plasma) membrane

2 Give **three** structural differences between a plant and animal cell.

2 A plant cell has a cellulose cell wall, chloroplasts, and a sap vacuole lined by a tonoplast.

Cell (plasma) **membrane** – gives physical protection to the outside of a cell, allows the import and export of selected chemicals.
Golgi body – is a packaging system where chemicals become membrane wrapped, forming vesicles
ribosomes – aid the manufacture of proteins, being the site where mRNA meets tRNA so that amino acids are bonded together
centrioles – help produce the spindle during cell division
mitochondria – release energy during aerobic respiration
1 **nucleus** – mRNA is produced in the nucleus with the help of DNA

2.2 Isolation of cell organelles

After studying this section you should be able to:

- understand how cell fractionation and ultracentrifugation are used to isolate cell organelles
- understand the principles of light and electron microscopes
- understand how microscopic specimens and microbial populations are measured

LEARNING SUMMARY

Cell fractionation

AQA A   M1
AQA B   M1

Occasionally it is necessary to isolate organelles to investigate their structure or function, e.g. mitochondria could be used to investigate aerobic respiration away from the cell's internal environment. **Cell fractionation** consists of two processes, **homogenisation** followed by **differential centrifugation**. Cell fractionation depends on the different densities of the organelles.

Technique

Cells are kept:

- cool at around 5°C *(this slows down the inevitable autolysis, destruction by the cell's own enzymes)*
- in an isotonic solution, i.e. equal concentration of substances inside and outside of the cell membrane *(this ensures that the organelles are not damaged by osmosis and can still function)*
- at a specific pH by a buffer solution *(this ensures that the organelles can still function, as they are kept in suitable conditions)*.

Homogenisation

The cells are homogenised in either a pestle homogeniser or blender. This breaks the cells up releasing the organelles and cytoplasm. Many organelles are not damaged at all by this process. At this stage there is a suspension of mixed organelles.

pestle homogeniser   blender

Differential centrifugation

Equal amounts of homogenised tissue samples are poured into the tubes of an **ultracentrifuge**. This instrument spins the cell contents at a force many times greater than gravity. Organelle separation by this technique is **density dependent**.

600 g means 600 times the force of gravity. If a plant cell was centrifuged at around this speed then chloroplasts would be contained in the sediment as well as nuclei.

First spin

The sample is spun at 600 g for 10 minutes. **Nuclei**, the organelles of greatest density, collect in the **sediment** at the base of the tube. All other cell contents are

The principle of pouring off the supernatant to leave the pure sediment behind can be repeated at the end of each spin. In this way the main organelles can be isolated.

Note that it is much easier to obtain nuclei, because they are isolated in the first spin. Ribosomes are isolated at the final spin.

found in the supernatant (the fluid above the sediment). Pouring off the supernatant leaves the sediment of relatively pure nuclei.

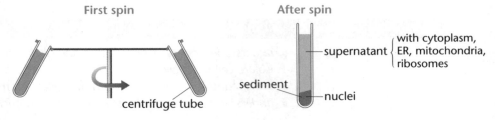

Second spin

The remaining supernatant fluid is spun at 10 000 to 20 000 g for a further 20 minutes. Mitochondria, the most dense of the remaining organelles collect in the sediment. All other cell contents are found in the supernatant.

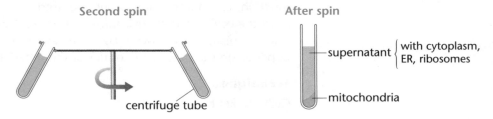

Third spin

The remaining supernatant fluid is spun at 100 000 g for a further 60 minutes. In this sediment are fragments of endoplasmic reticulum and ribosomes. Other cell contents e.g. cytoplasm and proteins remain in the supernatant.

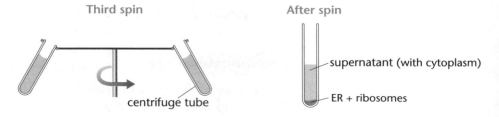

Using this technique the organelles are isolated. By providing them with suitable conditions they remain active for a time and can be used in investigations, e.g. chloroplasts given isotonic solution, suitable (warm) temperature, light, carbon dioxide and water, will continue to photosynthesise.

Electron and light microscopy

AQA A	M1
AQA B	M1
EDEXCEL	M1
OCR	M1
WJEC	M1
NICCEA	M1

Microscopes magnify the image of a specimen to enable the human eye to see minute objects not visible to the naked eye. Resolution of a microscope is the ability to distinguish between two objects as separate entities. At low resolution only one object may be detected. At high resolution two distinct objects are visible. At high resolution the image of such a specimen would show considerable detail.

The light microscope has limited resolution (0.02 µm) due to the wavelength of light so that organelles such as mitochondria, although visible, do not have clarity. Electron microscopes have exceptional resolution. The transmission electron microscope has a high resolution (0.2 to 0.3 nm). This enables even tiny organelles to be seen.

The light microscope

This type of microscope uses white light to illuminate a specimen. The light is focused onto the specimen by a condensing lens. The specimen is placed on a microscope slide which is clipped onto a platform, known as the stage. The image is viewed via an eyepiece or ocular lens. Overall magnification of the specimen depends on the individual magnification of the eyepiece lens and objective lens. For example, if a specimen is being observed with an eyepiece × 10 and an objective lens of × 40, then the image is 400 times the true size of the specimen.

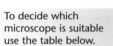

To decide which microscope is suitable use the table below.

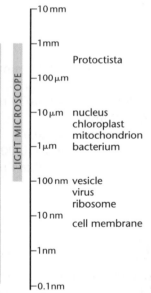

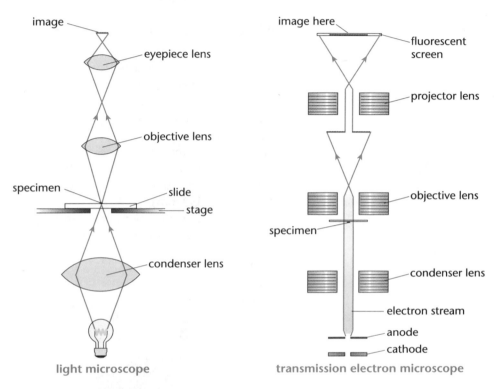

light microscope

transmission electron microscope

The electron microscope

This uses an electron stream which is directed at the specimen. The transmission electron microscope (TEM) has extremely high magnification and resolution properties. Specimens are placed in a vacuum within the microscope, to ensure the electrons do not collide with air molecules and distort the image. Stains such as osmium and uranium salts are used to make organelles distinct. These salts are absorbed by organelles and membranes differentially, e.g. the nuclear membrane absorbs more of the salts than other parts. In this way the nuclear membrane becomes more dense. When the electron beam hits the specimen, electrons are unable to pass through this dense membrane. The membrane shows up as a dark shadow area on the image, because it is in an electron shadow. Cytoplasm allows more electrons to pass through. When these electrons hit the fluorescent screen visible light is emitted.

Artefacts

When microscopic specimens are prepared there are often several chemical and physical procedures. Often the material is dead so changes from the living specimen are expected. Microscopic material should be analysed with care because there may have been some artificial change in the material during preparation, e.g. next to some cells a student sees a series of small circles. They look like eggs but are merely air bubbles. These are artefacts; structures alien to the material which should not be interpreted as part of the specimen.

Microscopic measurement

It is sometimes necessary to measure microscopic structures. There are two instruments needed for this process, a graticule and stage micrometer.

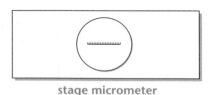

graticule   stage micrometer

Calibration and measurement technique

- Put a graticule into the eyepiece of a microscope
- look through the eyepiece lens and the graticule line can be seen
- put a stage micrometer on the microscope stage
- look through the eyepiece lens
- line up the ruled line of the graticule with the ruled line of the stage micrometer
- calibrate the eyepiece by finding out the number of eyepiece units (e.u.) equal to one stage unit (s.u.); each is 0.01 mm
- if three eyepiece units equal 1 stage unit, then 1 eyepiece unit is equal to 0.01/3 (0.0033 mm)
- take away the stage micrometer and replace it with a specimen
- measure the dimensions of the specimen in terms of eyepiece units.

Important! Every time you change the objective lens, e.g. move from low power to high power, recalibration is necessary. In an examination this may be tested. Many candidates forget to recalibrate. Don't miss the mark!

Example

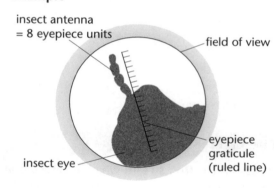

insect antenna = 8 eyepiece units

field of view

eyepiece graticule (ruled line)

insect eye

1 s.u. = 0.01 mm
3 e.u. = 1 s.u.
therefore 1 e.u. = 0.0033 mm
length of insect
    antenna = 8 e.u.
             = 8 × 0.0033 mm
             = 0.0264 mm

Estimation of populations

There are several techniques which can be used depending on the size of the cells or organisms. A haemocytometer can be used to estimate the numbers of red blood cells or minute organisms such as yeasts. A haemocytometer is a microscope slide on which an etched grid of squares has been drawn.

Technique:

- put a suspension of cells on the haemocytometer
- put a cover slip over the cell suspension to make sure that the depth is uniform (known depth, 0.1 mm)
- each small square is 0.0025 mm^2 in area and the volume above each square is 0.00025 mm^3
- look through the microscope and count the number of cells per square in a number of squares
- if a cell touches the sides of the square then only count vertical left and horizontal bottom cells
- calculate the average number of cells per square and then the number of cells per mm^3 or cm^3.

Important! In an examination you may be tested for your awareness of dilution. There are millions of red blood cells in even a small amount of blood (too many to count accurately). It is therefore always diluted before the haemocytometer count. You will need to multiply the amount of cells per mm^3 or cm^3 by the dilution factor (the question will supply this).

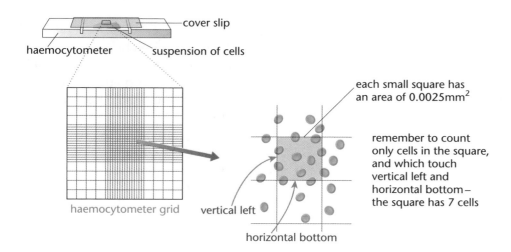

cover slip

haemocytometer   suspension of cells

each small square has an area of 0.0025mm^2

remember to count only cells in the square, and which touch vertical left and horizontal bottom – the square has 7 cells

haemocytometer grid

vertical left

horizontal bottom

Progress check

1 Describe how the width of an insect's leg could be measured using a microscope, graticule and stage micrometer.

2 A student measured the insect leg but considered that a higher magnification was needed to improve accuracy. What is the significance of changing magnification to the measuring technique?

1 Put a graticule into the eyepiece of a microscope; put a stage micrometer on the microscope stage; look through the eyepiece lens; line up the ruled line of the graticule with the ruled line of the stage micrometer; calibrate the eyepiece by finding out the number of eyepiece units equal to one stage unit; a stage unit is a known length, so an eyepiece unit length can be calculated; take away the stage micrometer and replace with specimen; measure width of insect leg in eyepiece units.

2 Recalibration is needed for each new magnification.

2.3 Specialisation of cells

After studying this section you should be able to:

- *understand cell specialisation and how cells aggregate into tissues and organs*
- *recall a range of cell adaptations*

The earlier parts of this chapter informed of the structure and function of generalised animal and plant cells. The described features are found in many unicellular organisms where all the life-giving processes are carried out in one cell. Additionally many multicellular organisms exist. A few show no specialisation and consist of repeated identical cells, e.g. Volvox, a colonial alga. Most multicellular organisms exhibit specialisation, where different cells are adapted for specific roles.

Cell adaptations

AQA A	M1
AQA B	M1
EDEXCEL	M1
OCR	M1
WJEC	M1
NICCEA	M1

Some important features:

Red blood cell
- no nucleus
- high surface area
- contains haemoglobin which has an affinity for oxygen

red blood cell

Endothelial cell
- very thin
- allows exchange of chemicals

endothelial cell

Plant palisade cell
- chloroplasts for photosynthesis
- chloroplasts can move to absorb more light
- contains chlorophyll which absorbs light
- sap vacuole stores important chemicals

Motor neurone
- can transmit electrical impulses
- has an insulative fatty sheath
- motor end plates to stimulate muscles to contract.

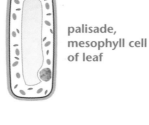

palisade, mesophyll cell of leaf

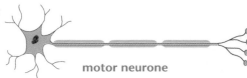

motor neurone

Tissues and organs

A **tissue** is a collection of similar cells, derived form the same source, all working together for a specific function, e.g. palisade cells of the leaf which photosynthesise or the smooth muscle cells of the intestine which carry out peristalsis.

An **organ** is a collection of tissues which combine their properties for a specific function, e.g. the stomach includes the tissues; smooth muscle, epithelial lining cells, connective tissue, etc. Together they enable the stomach to digest food.

A range of tissues and organs combine to form a system, e.g. the **respiratory system**.

In multicellular organisms specific groups of cells are specialised for a particular role. This increased efficiency helps the organism to have better survival qualities in the environment.

The photomicrograph below shows some of the cells which are part of a bone.

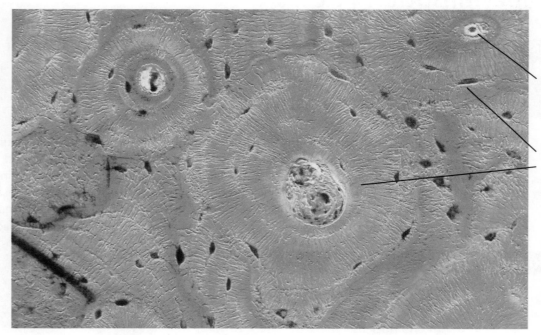

Haversian canal containing blood vessels and nerves

Bone cells which secrete the minerals which harden the bone

Combinations of cells each contribute their specific adaptations to the overall function of an organ. Compact bone, spongy bone and articular cartilage all have distinct but vital qualities.

Sample question and model answer

The electron micrograph below shows a lymphocyte which secretes antibodies. Antibodies are proteins.

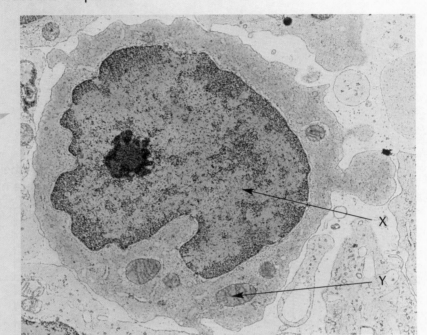

> Analyse electron micrographs carefully. Learn the typical structures of all the organelles then you will be prepared.

(a)

(i) Name the organelles X and Y. [2]

X = nucleus
Y = mitochondria

(ii) The cell was stained with uranium salts in preparation for a transmission electron microscope. Explain how this stain caused the nucleus to show a dark shade compared to the light shade of the cytoplasm. [4]

> There are many different stains. Most examination boards require that you know the principle of how they work, but not specific names.

The stain was taken up by the nucleus more than the cytoplasm; the electrons could not pass through the stained (dense) parts of the nucleus so the dark nucleus parts on the screen are in electron shade. Electrons pass through the cytoplasm and cause light emission (fluorescence) at the screen.

(b)

(i) Given a piece of liver how would you isolate mitochondria from the cells? [7]

> Note that there are 7 marks maximum for this question. Any seven of the responses shown would achieve maximum credit. The ideal answer given shows 9 potential creditable points.

Put the liver in isotonic solution;
homogenise the liver or grind with a pestle and mortar;
filter the homogenate through muslin layers to remove cell debris;
put in a centrifuge; spin at 500 – 600 g for 10 minutes;
discard the pellet or sediment; centrifuge the supernatant;
spin at 10 000 – 20 000 g for 20 minutes;
mitochondria are now in the pellet or sediment.

(ii) Why is it important to keep fresh liver cells at a temperature of around 5^{0}C during the preparation of the sample? [2]

They keep alive for a longer period since a low temperature slows down the action of the enzymes which break down the mitochondria; i.e. prevents or slows down cell autolysis (self-digestion of the cells).

Practice examination questions

1 The diagram shows the structure of a cell surface membrane.

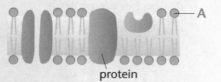

protein

(a) Name molecule A. [1]

(b) Describe the role of protein molecules in:
  (i) active transport
  (ii) facilitated diffusion. [4]

2 (a) Complete the table by ticking boxes to show the correct statement(s) for each method by which molecules cross a cell surface membrane.

	diffusion	facilitated diffusion	active transport
molecules move from where they are in high concentration to low concentration			
molecules move from where they are in low concentration to high concentration			
a protein carrier is needed			

[3]

(b) Name **two** other methods by which molecules can pass through the cell surface membrane. [2]

3 The diagram shows a section through a leaf.

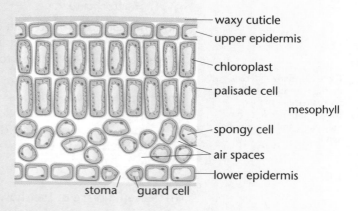

waxy cuticle
upper epidermis
chloroplast
palisade cell
mesophyll
spongy cell
air spaces
lower epidermis
stoma   guard cell

Explain how the leaf is adapted to achieve the maximum rate of photosynthesis. [7]

4 The diagram below shows a bacterium.

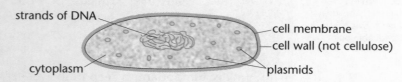

strands of DNA
cell membrane
cell wall (not cellulose)
cytoplasm
plasmids

(a) Describe **two** ways, visible in the diagram, which show that the bacterium is a prokaryotic organism. [2]

(b) Name **two** organelles from a human cell which show that it is a eukaryotic organism. [2]

Practice examination questions (continued)

5 Yeast population growth was investigated with the help of a haemocytometer. A sample from a yeast culture was diluted to 0.1 of the original concentration. The haemocytometer squares (below) show some of the yeast cells.

> Remember to use the dilution factor when calculating the original population.

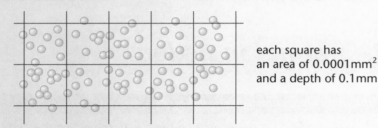

each square has an area of 0.0001mm² and a depth of 0.1mm

(a) (i)   Using the 10 haemocytometer squares in the diagram, calculate the average number of yeast cells per square.   [2]

(ii)   How many yeast cells would there be in 1 cm³ of the original yeast suspension? (1 cm³ = 1000 mm³)

Show your working.   [3]

(b) Why was it important to stir the yeast culture before putting the sample onto the haemocytometer?   [1]

6 This diagram below shows the structure of a transmission electron microscope (TEM).

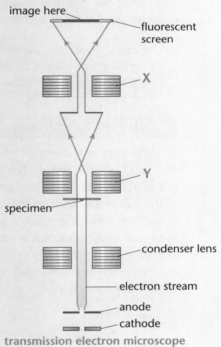

transmission electron microscope

(a) Name lens X and lens Y.   [2]

(b) Why is it necessary for the specimen to be put in a vacuum?   [1]

(c) Occasionally an image seen when using the electron microscope shows an item not present in the living organism.

(i)   What name is given to this type of item?   [1]

(ii) How should the presence of the item be interpreted?   [1]

Chapter 3
Enzymes

The following topics are covered in this chapter:

- *Enzymes in action*
- *Inhibition of enzymes*

- *Digestive enzymes*
- *Applications of enzymes*

3.1 Enzymes in action

After studying this section you should be able to:

LEARNING SUMMARY

- *understand the role of the active site and the enzyme–substrate complex in enzyme action*
- *understand how enzymes catalyse biochemical reactions by lowering activation energy*
- *understand the factors which affect the rate of enzyme catalysed reactions*

How enzymes work

AQA A	M1
AQA B	M1
EDEXCEL	M1
OCR	M1
WJEC	M1
NICCEA	M1

Living cells carry out many biochemical reactions. These reactions take place **rapidly** due to the presence of enzymes. All enzymes consist of **globular proteins** which have the ability to 'drive' biochemical reactions. Some enzymes require additional non-protein groups to enable them to work efficiently. The enzyme dehydrogenase needs a coenzyme NAD to function.

> The tertiary folding of polypeptides are responsible for the special shape of the active site.

KEY POINT

The ability of an enzyme to function depends on the specific shape of the protein molecule. The intricate shape created by polypeptide folding (see page 27) is a key factor in both theories of enzyme action.

Lock and key theory

> In an examination the lock and key theory is the most important model to consider. Remember that both catabolic and anabolic reactions may be given.

- Some part of the enzyme has a cavity with a precise shape (**active site**)
- a substrate can fit into the active site
- the active site (lock) is exactly the correct shape to fit the substrate (key)
- the substrate binds to the enzyme forming an **enzyme–substrate complex**
- the reaction takes place immediately
- certain enzymes break a substrate down into two or more products (**catabolic** reaction)
- other enzymes bond two or more substrates together to assemble one product (**anabolic** reaction).

> metabolic reaction
> = anabolic + catabolic
>   reaction   reaction
> Remember that metabolism is a summary of **build up** and **break down reactions**.

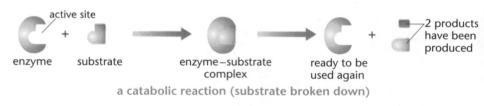

a catabolic reaction (substrate broken down)

an anabolic reaction (substrates used to build a new molecule)

Induced fit theory

- The active site is a cavity of a particular shape
- initially the active site is not the correct shape in which to fit the substrate
- as the substrate approaches the active site, the site changes and this results in it being a perfect fit
- after the reaction has taken place, and the products have gone, the active site returns to its normal shape.

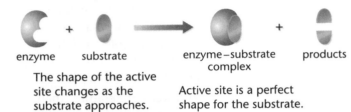

enzyme + substrate → enzyme–substrate complex + products

The shape of the active site changes as the substrate approaches.

Active site is a perfect shape for the substrate.

Lowering of activation energy

Every reaction requires the input of energy. Enzymes reduce the level of activation energy needed as shown by the graph.

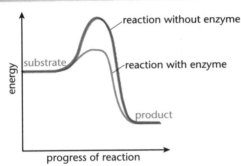

The higher the activation energy the slower the reaction. An enzyme reduces the amount of energy required for a biochemical reaction. When an enzyme binds with a substrate the available energy has a greater effect and the rate of catalysis increases. The conditions which exist during a reaction are very important when considering the rate of progress. Each of the following has an effect on the rate:

- concentration of substrate molecules
- concentration of enzyme molecules
- temperature
- pH.

You may be questioned on the factors which affect the rate of reaction. Less able candidates tend to remember just one or two factors. Learn all 4 factors here and achieve a higher grade!

What is the effect of enzyme concentration?

When considering the rate of an enzyme catalysed reaction the proportion of enzyme to substrate molecules should be considered. Every substrate molecule fits into an active site, then the reaction takes place. If there are more substrate molecules than enzyme molecules then the number of active sites available is a limiting factor. The optimum rate of reaction is achieved when all the active sites are in use. At this stage if more substrate is added, there is no increase in rate of product formation. When there are fewer substrate molecules than enzyme molecules the reaction will take place very quickly, as long as the conditions are appropriate.

Look out for questions which show the rate of reaction graphically. The examiners often test your understanding of limiting factors (see practice question).

How does temperature affect the rate of an enzyme catalysed reaction?

- Heat energy reaching the enzyme and substrate molecules causes them to increase random movement.
- The greater the heat energy the more the molecules move and so collide more often.
- The more collisions there are the greater the chance that substrates will fit into an active site, up to a specific temperature.

Remember that particles in liquids (and gases) are in constant random motion, even though we cannot see them.

- At the **optimum** temperature of an enzyme, the reaction rate is maximum.
- Heat energy also affects the shape of the active site, the active site being best at the optimum temperature.
- At temperatures above optimum, the rate of reaction decreases because the active site begins to distort.
- Very high temperature causes the enzyme to become **denatured**, i.e. bonding becomes irreversibly changed and the active site is **permanently damaged**.
- At very high temperatures, the number of collisions is correspondingly high, but without active sites no products can be formed.
- At lower temperatures than the optimum, the rate of the reaction decreases because of reduced enzyme/substrate collisions.

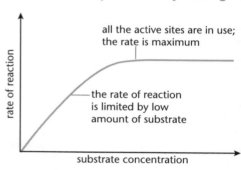

Most enzymes have an optimum between 30°C and 40°C, but there are many exceptions. An example of this is shown by some bacteria:

- **thermophiles** – enzymes optimum above 40°C
- **mesophiles** – enzymes optimum between 20°C and 40°C
- **psychrophiles** – enzymes efficient below 20°C.

> It is interesting to consider that some microbes can spoil ice cream in a freezer whereas a different microbe, with different enzymes, can decompose grass in a 'steaming' compost heap.

How does pH affect the rate of an enzyme catalysed reaction?

The **pH** of the medium can have a direct effect on the bonding responsible for the **secondary and tertiary structure** of enzymes. If the active site is changed then enzyme action will be affected. Each enzyme has an optimum pH.

- Many enzymes work best at **neutral** or **slightly alkaline** conditions, e.g. salivary amylase.
- Pepsin works best in **acid** conditions around pH 3.0, as expected considering that the stomach contains hydrochloric acid.

> Remember that other factors affect an enzyme catalysed reaction:
> - substrate concentration
> - enzyme concentration
> - temperature.
>
> Each can be a limiting factor.

For the two enzyme examples above, the active sites are ideally shaped at the pHs mentioned. An inappropriate pH, often acidic, can change the active site drastically, so that the substrate cannot bind. The reaction will not take place. On most occasions the change of shape is not permanent and can be returned to optimum by the addition of an alkali.

Progress check

How does temperature affect the rate of the reaction by which protein is changed to polypeptides by the enzyme pepsin, in the human stomach?

$$\text{protein} \xrightarrow{\text{pepsin}} \text{polypeptides}$$

- Heat energy causes the enzyme and substrate molecules to increase random movement, increasing the chance of collision.
- At 37°C (optimum temperature) there is a greater chance that the protein will fit into an active site, so the production of polypeptides is at maximum rate.
- At 37°C (optimum temperature) the shape of the active site is best suited to fit the protein.
- At temperatures higher than 37°C the rate of reaction decreases because the active site begins to distort.
- Very high temperature causes the pepsin to become denatured, i.e. bonding has been irreversibly changed and the active site is permanently damaged.
- At very high temperatures the number of collisions is correspondingly high, but without active sites no polypeptides can be formed.
- At lower temperatures than optimum the rate of reaction decreases because of reduced enzyme–substrate collisions.

3.2 Inhibition of enzymes

After studying this section you should be able to:

- *understand the action and effects of competitive and non-competitive inhibitors*
- *understand the process of end-product inhibition*

What are inhibitors?

AQA A	M1
AQA B	M1
EDEXCEL	M1
OCR	M1
WJEC	M1

If enzyme reactions inside the cell were to continue without regulation, there would be many problems. Cells possess regulatory chemicals which slow down or stop enzyme catalysed reactions. These chemicals are the inhibitors.

Competitive inhibitors

- These are molecules of similar shape to the normal substrate and are able to bind to the active site.
- They do not react within the active site, but leave after a time without any product forming.
- The enzymic reaction is reduced because while the inhibitor is in the active site, no substrate can enter.
- Substrate molecules compete for the active site so the rate of reaction decreases.
- The higher the proportion of competitive inhibitor the slower the rate of reaction.

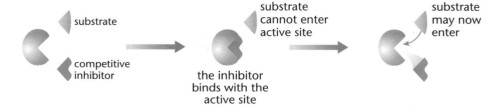

substrate / competitive inhibitor → the inhibitor binds with the active site / substrate cannot enter active site → substrate may now enter

Some enzymes have two sites, the active site and one other. An **allosteric** molecule fits into the alternative site. Here it changes the shape of the active site. This can stimulate the reaction if the active site becomes a better shape (**allosteric activation**). It can also inhibit if the active site becomes an inappropriate shape (**allosteric inhibition**).

Non-competitive inhibitors

- These are molecules which bind to some part of an enzyme other than the active site.
- They have a different shape to the normal substrate.
- They change the shape of the active site which no longer allows binding of the substrate.
- Some substrate molecules may reach the active site before the non-competitive inhibitor.
- The rate of reaction is reduced.
- Finally they leave their binding sites, but substrate molecules do not compete for these, so they have a greater inhibitory effect.

non-competitive inhibitor / substrate / binding site → active site has changed → substrate has opportunity to enter

The graph below shows the relative effects of competitive and non-competitive inhibitors, compared to a normal enzyme catalysed reaction.

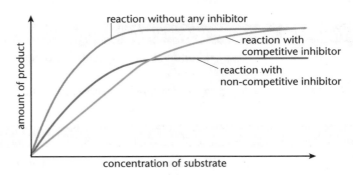

End-product inhibition

This mechanism is needed to regulate certain enzyme catalysed processes in organisms. It involves allosteric sites. The diagram below shows one example of end-product inhibition.

Stage 1
A substrate binds with the active site of enzyme X. A product is formed.

Stage 2
This product then binds with the active site of enzyme Y. Another product is formed.

Stage 3
The stage 2 product then binds with the active site of enzyme Z. Another product is formed.

Stage 4
This final product is the feedback product. It is the correct shape to bind with the allosteric site of enzyme X. Once in position it distorts the active site, inhibiting the first reaction and those which follow.
The final end product has caused its own decrease.

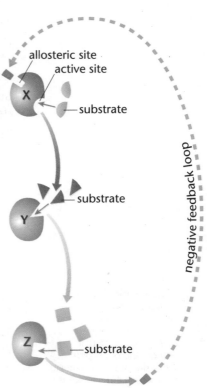

Progress check

1 Does the substrate compete with a non-competitive inhibitor to bind with the active site of an enzyme? Give a reason for your answer.

2 How does a non-competitive inhibitor reduce the rate of an enzyme catalysed reaction?

1 No. Non-competitive inhibitors have a different shape to the normal substrate. They bind to some part of an enzyme other than the active site.
2 They change the shape of the enzyme's active site which is less suitable for the binding of the substrate, so the rate of reaction decreases.

3.3 Digestive enzymes

After studying this section you should be able to:

- recall the sites of secretion and action of a range of human digestive enzymes
- recall extracellular digestion by saprobiotic microorganisms

Human digestive enzymes

AQA B M1
EDEXCEL M2
NICCEA M1

Animals which feed heterotrophically take in complex organic substances. These are subjected to physical breakdown, giving enzymes a more exposed surface area on which to act. The human digestive enzymes are hydrolases (see page 23). Each time a molecule is broken down water is required during the reaction. The table below includes some human digestive enzymes.

Lipase is aided by bile, an alkaline secretion. It reaches the duodenum and emulsifies lipids, breaking them up into tiny globules of lipid (a high surface area for the lipase to act on). It helps to neutralise acid from the stomach so that enzymes in the duodenum and small intestine have a suitable environment.

site in alimentary canal	secretion	enzyme	substrate	product
mouth	saliva	amylase	starch	maltose
stomach	gastric juice	pepsin	protein	polypeptides
duodenum	pancreatic juice	lipase	lipids	fatty acid glycerol
duodenum	pancreatic juice	amylase	starch	maltose
small intestine	intestinal juice	maltase	maltose	glucose

KEY POINT

Pepsin is an endopeptidase. This means that it breaks peptide bonds in the middle of a polypeptide chain. Exopeptidases break peptide bonds to remove individual amino acids from the ends of polypeptides.

These two types of protease work together, the endopeptidases producing shorter polypeptide chains, consequently the exopeptidases have more ends to work on.

Saprobiotic bacteria and fungi

AQA A M1
AQA B M1
EDEXCEL M1
OCR M1
WJEC M1
NICCEA M1

These microorganisms secrete extracellular enzymes to obtain nutrients by decaying organic matter. The enzymes act outside the cell, so that smaller nutrient molecules can be absorbed in solution. The diagram below shows a saprobiotic fungus decaying a starch-rich potato.

Most examination boards use the term saprobiotic instead of saprophytic. Both refer to decomposers.

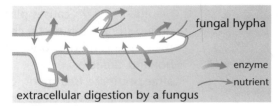

extracellular digestion by a fungus

Progress check

The diagram shows a saprobiotic fungus feeding on starch.

Describe and explain how the fungus is able to obtain a supply of soluble glucose.

The fungus secretes an extracellular enzyme, amylase. This breaks down starch into maltose. It also secretes maltase which breaks down maltose into glucose. These are soluble compounds which can be absorbed through the fungal cell membrane by the process of diffusion.

3.4 Applications of enzymes

After studying this section you should be able to:

- *describe and understand a range of home, medical and industrial applications of enzymes*
- *understand the advantages of immobilised enzymes*

Home, medical and industrial applications of enzymes

AQA A ▸ M1
AQA B ▸ M1
EDEXCEL ▸ M1
WJEC ▸ M1
NICCEA ▸ M1

There are a wide range of applications of enzymes used in processes throughout the world. Historically enzymes have been exploited for many years, e.g. yeast cells are still used in fermentation to produce ethanol. The yeast is really used as an enzyme 'package' to change sugar into ethanol via a sequence of enzyme catalysed reactions.

We currently live in a golden age of biotechnology and will witness many new applications of enzyme use within the coming years.

Biological detergents

These are products such as biological washing and dishwasher powders. The common link between them is they contain a range of hydrolysing enzymes. Manufacturers rarely inform the public of the specific enzymes in their detergents, but invariably they are:

- amylases – break down starch stains
- cellulases – break down the ends of damaged cotton fibres to remove the 'fuzz' produced during washing
- lipase – breaks down lipid stains into fatty acids and glycerol
- proteases break down the many different proteins found in food stains.

Note the pattern – enzymes often end in -ase.

This enzyme 'cocktail' has a low temperature optimum around 50°C, so that much less electricity is needed for washing; additionally, difficult stains are removed.

The enzymes are produced in fermenters (see page 88).

Fruit juice extraction

Crushing fruit such as apples releases juice, a valuable food product. Extraction of the juice is aided by an enzyme as follows :

- insoluble pectin causes plant cell walls to adhere to adjacent cells
- during storage this pectin changes to a soluble form which binds water strongly
- pectinase is used to break down the pectin chains which reduces its water-holding capacity
- after pectinase treatment, crushing releases a greater yield of juice
- the pectinase even clears the juice of 'cloudiness' caused by pectin.

You could be given a question about a different enzyme which you have not studied before. Do not panic! Enough information will be supplied and all you need to do is apply your knowledge of enzyme principles.

Lactose removal from milk

This is necessary because some people have allergic reactions to lactose, the sugar in milk. Lactase is added to change the lactose to glucose and galactose. As a result, lactose-free milk is available to consumers.

Clinistix

These are strips of cellulose which have the enzyme glucose oxidase stuck to one end. When dipped into urine containing glucose the reaction produces hydrogen

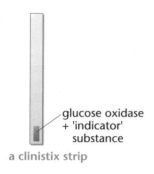

glucose oxidase + 'indicator' substance

a clinistix strip

peroxide. This reacts with another compound to give a colour change. This colour change signals that the patient is potentially diabetic.

Biosensors

These are devices which are used to detect specific molecules. All biosensors use a biological component such as a layer of enzymes or antibodies. These produce a signal in the presence of a specific chemical. They work as follows:

- a sample is placed in contact with the biosensor
- when enzymes are used in the 'detection membrane', if the chemical being investigated is present, the molecules fit into the active sites of the enzyme
- this causes a transducer to produce an electrical signal
- the level of electrical signal is proportional to the level of chemical, so the device is quantitative.

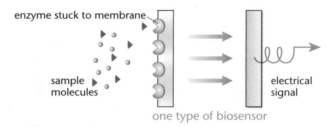

enzyme stuck to membrane

sample molecules

electrical signal

one type of biosensor

ELISA (Enzyme-linked immunosorbent assay)

This is used to detect minute quantities of protein in fluids such as blood plasma. It works as follows:

- if protein X is to be detected
- antibodies known to bind to protein X, are fixed to a plate
- protein X, *if present*, binds to the antibody whereas all other proteins wash away
- a second type of antibody, linked to an enzyme, is added and binds to X at a different site to the first antibody
- a specific substrate is added which binds to the active site of this enzyme
- the product formed is easily detectable, e.g. brightly coloured
- this product would not form in the plate at all if protein X was absent.

Some diseases can be diagnosed by ELISAs, as specific proteins are detected in blood plasma. This technique has many applications including the detection of pregnancy, bacteria and viruses.

Immobilised enzymes

After completion of an enzyme catalysed reaction the enzyme remains unchanged and can be used again. Unfortunately the enzyme can contaminate the product, as they can be difficult to separate out from the reaction mixture. For this reason immobilised enzymes have been developed. They are used as follows:

- enzymes are attached to insoluble substances such as resins and alginates
- these substances usually form membranes or beads and the enzymes bind to the outside
- substrate molecules readily bind with the active sites and the normal reactions go ahead
- the immobilised enzymes are easy to recover, remaining in the membranes or beads
- there is no contamination of the product by free enzymes
- expensive enzymes are re-used
- processes can be continuous unlike batch, where the process is stopped for 'harvesting'.

Consider the methods of immobilised enzymes. The more bullet points you remember, the more marks you will gain. Top exam candidates can recall most points. Look at the number of marks available at the end of a question. This informs you of the number of creditable points you need to make to score highly!

Sample questions and model answers

1

The graph below shows the rate of an enzyme catalysed reaction, with and without a **non-competitive inhibitor**, at different substrate concentrations.

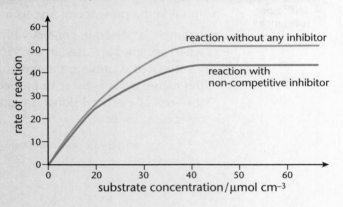

(a) Give **one** piece of evidence from the graph which shows that the inhibitor was non-competitive rather than competitive. [1]

When the non-competitive inhibitor is present the maximum rate of the inhibitor-free reaction is not reached.

> A competitive inhibitor would still allow the maximum rate at high substrate concentration.

(b) Explain why the rates of the inhibited and non-inhibited reactions were very similar up to a substrate concentration of 20 μmol cm⁻³. [2]

Before a concentration of 20 μmol cm⁻³
- in both reactions substrate molecules are similarly successful in reaching active sites
- few inhibitor molecules have become bound to the alternative sites on the enzyme
- few active sites have been changed so are still available for the substrate.

> Remember that the non-competitive inhibitor binds to a different part of an enzyme and NOT the active site. It still causes a change in the active site which cannot then bind with the substrate.

(c) Give **two** factors that are needed to be kept constant when investigating both the inhibited and non-inhibited reactions. [2]

- temperature
- pH

2

The sequence below shows a biochemical pathway which involves three enzymes E_1, E_2 and E_3 to form molecule X.

> Be ready to use any visual clues about the active sites and *alternative* sites.

(a) Explain how product X is an allosteric inhibitor of E_1 [2]

The molecule fits into an alternative site on E_1 and the shape of active site is changed, the substrate therefore cannot bind to active site.

(b) Name the type of inhibition shown by this sequence. [1]

End-product inhibition.

Practice examination questions

1. The diagram below shows a biosensor which can detect very small quantities of urea in blood and urine.

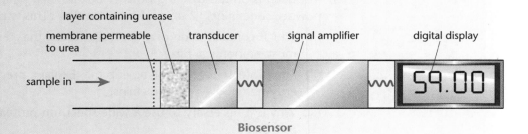

Biosensor

(a) Explain how the biosensor shown can be used to detect a minute quantity of urea molecules in a sample of blood. [3]

(b) How is the biosensor able to detect different amounts of urea in each sample? [1]

2. Describe each of the following pairs to show that you understand the main differences between them:

(a) the lock and key enzyme theory **and** the induced fit enzyme theory

(b) endopeptidases **and** exopeptidases. [4]

3. In an industrial process silver is reclaimed from a waste fluid. The silver is held on cellulose film by protein. An expensive enzyme is used to remove the protein. It is immobilised in a gel layer as fluid is passed over it.

State the advantages of using an immobilised enzyme in this process. [2]

4. In terms of the tertiary structure of the enzyme, explain why amylase can break down starch but has no effect on a lipid. [3]

5. Bacterial α-amylase works best at around 80°C.

(a) Name the substrate which it breaks down. [1]

(b) Why is this enzyme described as a thermostable enzyme? [1]

6. *Biox* is a biological washing powder which contains a wide spectrum protease. The graph shows the effect of a range of concentrations of protease on the removal of protein stains.

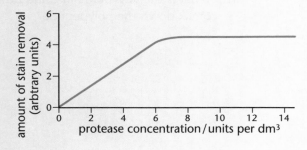

(a) What is the effect of the concentration of protease on stain removal between 2 and 6 units of protease? [1]

(b) Instructions on the washing powder box advised people to use an amount of powder containing 12 units of protease dm^{-2}. This was not good advice.

   (i) Which concentration of protease removed the stains at the fastest and most economical rate? [1]

   (ii) The protease is a hydrolytic enzyme. Explain, precisely, how this enzyme removes protein stains. [3]

   (iii) Why is it necessary to use a wide-spectrum protease in the washing powder? [1]

   (iv) Name a molecule which will be found in the waste water as a result of protein breakdown. [1]

(c) Name **two** other types of enzyme that would remove different stain components. [2]

7 The diagram represents an enzyme and its substrate.

(a) Referring to information in the diagram explain the activity of this enzyme in terms of the **induced fit theory**. [2]

(b) Molecule X is a non-competitive inhibitor. Explain how this inhibitor has an effect on the structure and function of the enzyme. [3]

8 The diagram below shows a long polypeptide.

(a) The carboxylic acid is found at one end of the polypeptide. Which group is found at the other end? [1]

(b) What is the advantage of using both an exopeptidase and an endopeptidase to break down the polypeptide? [3]

Chapter 4

Exchange

The following topics are covered in this chapter:

- The cell surface membrane
- The movement of molecules in and out of cells
- Gaseous exchange

4.1 The cell surface membrane

After studying this section you should be able to:

- understand the importance of surface area to volume ratio
- recall the fluid mosaic model of the cell surface (plasma) membrane

How important is the surface area of exchange surfaces?

AQA A	M1
AQA B	M1
EDEXCEL	M1
OCR	M1
WJEC	M1
NICCEA	M1

Unicellular organisms like amoeba have a very high surface area to volume ratio. All chemicals needed can pass into the cells directly and all waste can pass out efficiently. Organisms which have a high surface area to volume ratio have no need for special structures like lungs or gills.

Nutrients and oxygen passing into an organism are rapidly used up. This gives a limit on the ultimate size to which a microorganism can grow. If vital chemicals did not reach all parts of a cell then death would be a consequence.

A unicellular organism may satisfy all its needs by direct diffusion. However, in larger organisms cells join to adjacent ones, surfaces exposed for exchange of substances are reduced. The larger an organism the lower is its surface area to volume ratio. For this reason many multicellular organisms have specially adapted exchange structures.

Fluid mosaic model of the cell surface (plasma) membrane

AQA A	M1
AQA B	M1
EDEXCEL	M1
OCR	M1
WJEC	M1
NICCEA	M1

Remember that plant cells have a cellulose cell wall. This gives physical support to the cell but is permeable to many molecules. Water and ions can readily pass through.

Ultimately the exchange of substances takes place across the cell surface membrane. This must be selective, allowing some substances in and excluding others. The cell membrane consists of a bilayer of phospholipid molecules (see page 25). Each phospholipid is arranged so that the hydrophilic head (attracts water) is facing towards either the cytoplasm or the outside of the cell. The hydrophobic (repels water) tails meet in the middle of the membrane. Across this expanse of phospholipids are a number of protein molecules. Some of the proteins (intrinsic) span the complete width of the membrane, some proteins (extrinsic) are partially embedded in the membrane.

The fluid mosaic model of the cell membrane

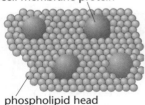

upper surface of cell membrane protein

phospholipid head

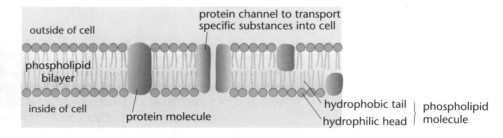

outside of cell

phospholipid bilayer

inside of cell

protein molecule

protein channel to transport specific substances into cell

hydrophobic tail
hydrophilic head
} phospholipid molecule

Functions of cell membrane molecules

The term 'fluid mosaic' was given because of the dynamic nature of the component molecules of the membrane. Many of the proteins seem to 'float' through an apparent 'sea' of phospholipids. Few molecules are static.

Phospholipid

Small lipid-soluble molecules pass through the membrane easily because they dissolve as they pass through the phospholipid bilayer. Small uncharged molecules also pass through the bilayer.

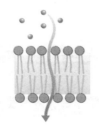

small lipid-soluble molecules pass through

Channel proteins (ion gates)

Larger molecules and charged molecules can pass through the membrane due to channel proteins. Some are adjacent to a receptor protein, e.g. at a synapse a transmitter substance binds to a receptor protein. This opens the channel protein or ion gate and sodium ions flow in.

Not all channel proteins need a receptor protein.

> When the molecule binds to a receptor molecule it is similar to a substrate binding with an enzyme's active site. On this occasion the receptor site is the correct shape.

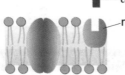

transmitter substance
receptor protein

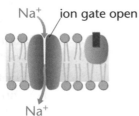

Na^+ ion gate open

Na^+

Carrier protein molecule

Some molecules which approach a cell may bind with a carrier protein. This has a site which the incoming molecule can bind to. This causes a change of shape in the carrier protein which deposits the molecule into the cell cytoplasm.

once in position the molecule changes the shape of the carrier protein

the site gives up the molecule on the inside of the cell

carrier protein

Recognition proteins

These are extrinsic proteins, some having carbohydrate components, which help in cell recognition and cell interaction, e.g. foreign protein on a bacterium would be recognised by white blood cells and the cell would be attacked.

> White blood cells continually check the proteins on cell membranes. Those recognised as 'self' are not attacked, whereas those which are not 'self' are attacked.

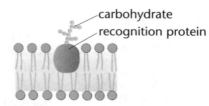

carbohydrate
recognition protein

> **KEY POINT**
>
> The cell surface membrane is the key structure which forms a barrier between the cell and its environment. Nutrients, water and ions must enter and waste molecules must leave. Equally important is the exclusion of dangerous chemicals and inclusion of vital cell contents. It is no surprise that the cell makes further compartments within the cell using membranes of similar structure to the cell surface membrane.

4.2 The movement of molecules in and out of cells

After studying this section you should be able to:

- *understand the range of methods by which molecules cross cell membranes*
- *understand the processes of diffusion, facilitated diffusion, osmosis and active transport*

How do substances cross the cell surface membrane?

AQA A	M1
AQA B	M1
EDEXCEL	M1
OCR	M1
WJEC	M1
NICCEA	M1

Cells need to obtain substances vital in sustaining life. Some cells secrete useful substances but all cells excrete waste substances. There are several mechanisms by which molecules move across the cell surface membrane.

Diffusion

> Note that diffusion is the movement of molecules down a concentration gradient.

Molecules in liquids and gases are in constant random motion. When different concentrations are in contact, the molecules move so that they are in equal concentration throughout. An example of this is when sugar is put into a cup of tea. If left, sugar molecules will distribute themselves evenly, even without stirring. Diffusion is the movement of molecules from where they are in high concentration to where they are in low concentration. Once evenly distributed the *net* movement of molecules stops.

Factors which affect the rate of diffusion

- surface area
  the greater the surface area the greater the rate of diffusion
- the difference in concentration at either side of the membrane
  the greater the difference the greater the rate
- the size of molecules
  smaller molecules may pass through the membrane faster than larger ones
- the presence of pores in the membrane
  pores can speed up diffusion
- the width of the membrane
  the thinner the membrane the faster the rate.

Facilitated diffusion

> Sometimes the membrane is stated as being selectively permeable, partially permeable or semi-permeable. It depends on which Examination board sets your papers.

This is a special form of diffusion in which protein carrier molecules (see page 58) are involved. It is much faster than regular diffusion because of the carrier molecules. Each carrier will only bind with a specific molecule. Binding changes the shape of the carrier which then deposits the molecule into the cytoplasm. No energy is used in the process.

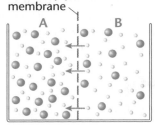

selectively permeable membrane

- water molecule
- solute molecule

water molecules move from B to A

Osmosis

This is the movement of water molecules across a selectively permeable membrane:

- from a lower concentrated solution to a higher concentrated solution
- from where water molecules are at a higher concentration to where they are at a lower concentration
- from a hypotonic solution to a hypertonic solution
- from a hyperosmotic solution to a hypo-osmotic solution
- from an area of higher water potential to lower water potential.

> Remember that osmosis is about the movement of **water molecules**. No other substance moves!

The diagram on the left shows a model of osmosis.

What is the relationship between water potential of the cell and the concentration of an external solution?

The term water potential is used as a measure of water movement from one place to another in a plant. It is measured in terms of pressure and the units are either kPa (kilopascals) or MPa (megapascals). Water potential is indicated by the symbol ψ *(pronounced psi)*. The following equation allows us to work out the water 'status' of a plant cell.

ψ (cell) water potential (of cell)	=	ψs solute potential (of ions inside cell)	+	ψp pressure potential (of cell wall)	**K E Y** **P O I N T**

Remember that water moves from an area of higher water potential to an area of lower water potential. When a cell at -4 MPa is next to a cell at the less negative value the water moves to the more negative value, i.e. -4 MPa > -6 MPa

Note that pressure potential only has a value **above** zero when the cell membrane **begins** to contact the cell wall. The greater the pressure potential the more the cell wall resists water entry. At turgidity $\psi s = \psi p$ when net water movement is zero.

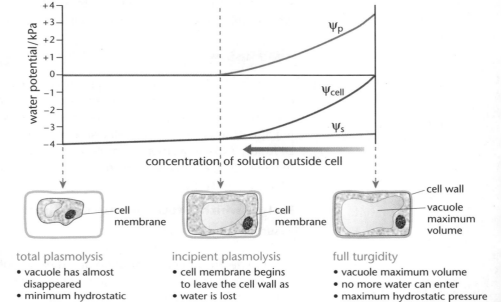

total plasmolysis
- vacuole has almost disappeared
- minimum hydrostatic pressure
- also known as flaccid

incipient plasmolysis
- cell membrane begins to leave the cell wall as
- water is lost

full turgidity
- vacuole maximum volume
- no more water can enter
- maximum hydrostatic pressure
- cell membrane is forced against the cell wall

Active transport

Note that active transport is the movement of molecules up a concentration gradient.

In **active transport** molecules move from where they are in lower concentration to where they are in higher concentration. A protein carrier molecule is used (see page 58). This is **against the concentration gradient** and always **needs energy**. A plant may contain a higher concentration of Mg^{2+} ions than the soil. It obtains a supply by active transport through the cell surface membranes of root hairs. Only Mg^{2+} ions can bind with the specific protein carrier molecules responsible for their entry into the plant. This is also known as active ion uptake, but is a form of active transport.

Endocytosis, exocytosis and pinocytosis

endocytosis

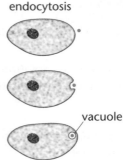

vacuole

Some substances, often due to their large size, enter cells by **endocytosis** as follows:

- the substance contacts the cell surface membrane which indents
- the substance is surrounded by the membrane, forming a vacuole or vesicle
- each vacuole contains the substance and an outer membrane which has detached from the cell surface membrane.

When fluids enter the cell in this way this is known as **pinocytosis**. Some substances leave the cell in a reverse of endocytosis. Here the membrane of the vacuole or vesicle merges with the cell surface membrane depositing its contents into the outside environment of the cell. This is known as **exocytosis**.

Progress check

1 A plant contains a greater concentration of Fe^{2+} ions than the soil in which it is growing. Name and describe the process by which the plant absorbs the ions against the concentration gradient.

2 Explain the following:
(a) Endocytosis of an antigen by a phagocyte
(b) Exocytosis of acetylcholine molecules from a cell.

acetylcholine contents deposited outside of the cell.
(b) Exocytosis: a vesicle in the cell contains acetylcholine molecules; the vesicle merges with the cell membrane;
vacuole; the vacuole contains the antigen which has detached from the cell surface membrane.
2 (a) Endocytosis: antigen contacts the cell membrane of the phagocyte; cell membrane surrounds the antigen, forming a
which allow entry into the plant.
root hairs; protein carrier molecules in membranes used; energy needed; Fe^{2+} ions can bind with the protein carrier molecules
1 Active transport: molecules move from a lower concentration to a higher concentration; through the cell surface membranes of

4.3 Gaseous exchange

After studying this section you should be able to:

• understand gaseous exchange in a dicotyledonous leaf, the gills of a bony fish and the lungs of a mammal
• show awareness of the adaptations of leaves, gills and lungs for efficient gaseous exchange

LEARNING SUMMARY

How are organisms adapted for efficient gaseous exchange?

AQA A	M1
AQA B	M1
EDEXCEL	M1
OCR	M1
WJEC	M2
NICCEA	M2

The range of respiratory surfaces in this chapter each have common properties, such as high surface area to volume ratio, one cell thick lining tissue, many capillaries.

Remember that all cells without chloroplasts must be supplied by cells capable of photosynthesis.

The exchange of substances across cell surface membranes has been described. Larger organisms have a major problem in exchange because of their low surface area to volume ratio. They satisfy their needs by having tissues and organs which have special adaptations for efficient exchange. In simple terms, these structures achieve a very high surface area, e.g. a leaf, and link to the transport system to allow import and export from the organ.

A dicotyledonous leaf

The diagram below shows a section through a leaf. Leaves of plants give a high surface area over which exchange takes place. Specialised tissues increase the efficiency of exchange to allow photosynthesis to supply the plant with *enough* energy-rich carbohydrates.

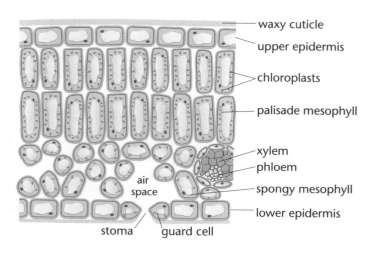

61

Adaptations of a leaf for photosynthesis

- A flat, thin blade (lamina) to allow maximum light absorption.
- Cells of the upper epidermis have a waxy cuticle to reflect excess light but allow entry of enough light for photosynthesis.
- Each leaf has many chloroplasts to absorb a maximum amount of light.
- Chloroplasts contain many thylakoid membranes, stacked in grana to give a high surface area to absorb the maximum quantity of light.
- Palisade cells, containing chloroplasts, pack closely together to 'capture' the maximum amount of light.
- Many guard cells open stomata to allow carbon dioxide in and oxygen out during photosynthesis.
- Air spaces in the mesophyll store lots of carbon dioxide for photosynthesis or lots of oxygen for respiration.
- Xylem of the vascular bundles brings water to the leaf for photosynthesis.
- Phloem takes the carbohydrate away from the leaf after photosynthesis.

Note that the leaf gives off oxygen during the day whilst the leaf is photosynthesising but gives off carbon dioxide at night during dark conditions when only respiration takes place.

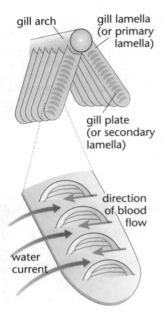

gill arch
gill lamella (or primary lamella)
gill plate (or secondary lamella)
direction of blood flow
water current

Gills of a bony fish

The ventilation mechanism of a fish allows intake of water, and passes it across the gills. The diagram (left) shows the structures of the gills which allow maximum exchange to take place.

Adaptations of gills for gaseous exchange

- The gills of a bony fish have a very high surface area to volume ratio.
- Gills consist of many flat gill filaments, stacked on top of each other, to give a high surface area for maximum exchange.
- Each gill filament has many gill plates which further increase surface area.
- Gill plates are very thin and full of blood capillaries to aid exchange.
- The gradients of O_2 and CO_2 are kept at a maximum by the counter-current flow mechanism. By allowing water to flow over the gills in an opposite direction to blood, maximum diffusion rate is achieved.

Remember that all respiratory surfaces are damp to allow effective transport across cell membranes.

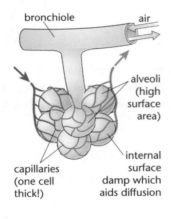

bronchiole
air
alveoli (high surface area)
internal surface damp which aids diffusion
capillaries (one cell thick!)

Lungs of a mammal

The ventilation mechanism of a mammal allows inhalation of air, which is passed into alveoli to exchange the respiratory gases. Completion of ventilation takes place when gases are expelled into the atmosphere. The diagram on the left shows the structure of alveoli.

Adaptations of lungs for gaseous exchange

- Air flows through a trachea (windpipe) supported by cartilage.
- It reaches the alveoli via tubes known as bronchi and bronchioles.
- Lungs have many alveoli (air sacs) which have a high surface area.
- Each alveolus is very thin (diffusion is faster over a short distance).
- Each alveolus has many capillaries, each one cell thick, to aid diffusion.
- There are many blood vessels in the lungs to give a high surface area for gaseous exchange and transport of respiratory substances.

Progress check

Describe and explain how the gills of a bony fish are adapted for efficient gaseous exchange.

Each gill consists of many thin gill filaments, stacked on top of each other which give a high surface area to volume ratio, for maximum exchange of gases; each gill filament has many gill plates which further increase surface area; each gill plate is very thin and full of blood capillaries; the gradients of O_2 and CO_2 are kept at a maximum by counter-current flow; water flows over the gills in an opposite direction to blood to achieve maximum diffusion rates.

Sample question and model answer

The diagram below shows two adjacent plant cells A and B.
The water potential equation is:

$$\psi(cell) = \psi s + \psi p$$

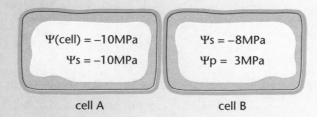

cell A   cell B

(a) (i) Calculate the water potential of cell B.   [1]

> Using the equation:
> $\psi(cell) = \psi s + \psi p$
> $= -8\ MPa + 3\ MPa$
> $= -5\ MPa$

In this question you were given the equation. Try to remember it because it could earn you a mark. Additionally, if there were calculations you need the equation to access the other marks.

When given any two of the values you can work out the other.

(ii) Draw an arrow on the diagram to show the direction of water flow. Show how you worked out the direction.   [2]

> The arrow should be drawn from cell B to cell A.
> Direction from -5 MPa to -10 MPa.

(iii) What is the value of the pressure potential (ψp) of cell A?   [1]

> $\psi(cell) = \psi s + \psi p,$
> $-10\ MPa = -10\ MPa + \psi p$
> $\psi p = -10\ MPa + 10\ MPa$
> $= 0\ MPa$

Note that from total plasmolysis up to incipient plasmolysis the resistance of the cell, i.e. ψp is zero.

Only when the cell membrane contacts with the wall does it have an effect.

(iv) Name the condition of the cell when $\psi(cell) = 0$ [1]

> full turgor or fully turgid

(b) Give **one** difference between the following terms: [2]

   facilitated diffusion

> molecules move down a gradient

   active transport

> energy is needed for the process

Be careful with this type of question. You may believe that 'up a gradient' could be given for active transport. It is correct, but it's too close to the 'down a gradient' idea for facilitated diffusion.

Go for a completely different idea, as shown.

(c) What effect would the following temperatures have on the active transport of Mg^{2+} ions across a cell surface membrane of a plant cell? Assume the plant is a British native.   [4]

(i) 30°C

> It is likely that active transport would be efficient because the temperature would be ideal for the Mg^{2+} to bind with a carrier molecule.

(ii) 80°C

> process likely not to work;
> protein carrier denatured;
> Mg^{2+} would not be able to bind.

Practice examination questions

1

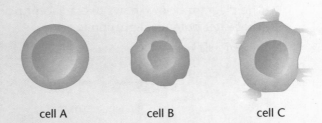

cell A   cell B   cell C

(a) Explain each of the following in terms of water potential.

  (i) Cell A did not change size at all.

  (ii) Cell B decreased in volume.

  (iii) Cell C became swollen and burst.   [3]

(b) Which process is responsible for the changes to cells B and C?   [1]

2 (a) Give one similarity between active transport and facilitated diffusion.   [1]

  (b) Give one difference between active transport and facilitated diffusion.   [1]

3 The diagram shows a section through a leaf.

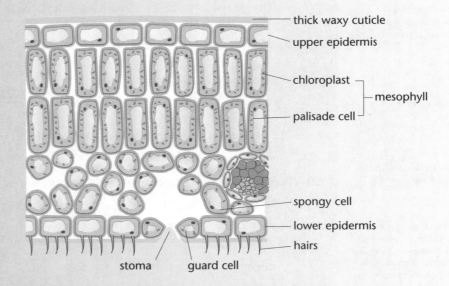

(a) Describe how the leaf is adapted for efficient photosynthesis.   [6]

(b) Suggest how the leaf is adapted to xerophytic conditions.   [3]

4 Describe and explain how the following are adapted to efficient gaseous exchange.

  (a) The alveoli of lungs.   [4]

  (b) The gills of a fish.   [4]

Chapter 5
Transport

The following topics are covered in this chapter:

- *The importance of mass transport systems*
- *Heart: structure and function*
- *Blood vessels*
- *The transport of substances in the blood*
- *The transport of substances in a plant*
- *Water loss in a plant*

5.1 The importance of mass transport systems

After studying this section you should be able to:

LEARNING SUMMARY

- *explain why most multicellular organisms need a mass transport system*
- *understand the importance of a high surface area to volume ratio*

Why do most multicellular organisms need a mass transport system?

AQA A	M1
AQA B	M3
EDEXCEL	M2
OCR	M3
WJEC	M2
NICCEA	M2

The bigger an organism is, the lower its surface area to volume ratio. Substances needed by a large organism could not be supplied through its exposed external surface. Oxygen passing through an external surface would be rapidly used up before reaching the many layers of underlying cells. Similarly waste substances would not be excreted quickly enough. This problem has been solved, through evolution, by specially adapted tissues and organs.

> Leaves, roots, gills and lungs all have high surface area to volume properties so that supplies of substances vital to **all** the living cells are made available by these structures. Movement of substances to and from these structures is carried out by efficient mass transport systems.
>
> **KEY POINT**

Across the range of multicellular organisms found in the living world are a number of mass transport systems, e.g. the mammalian circulatory system and the vascular system of a plant.

Mass transport systems are just as important for the rapid removal of waste as they are for supplies. Supplies include an immense number of substances, e.g. glucose, oxygen and ions. Even communication from one cell to another can take place via a mass transport system, e.g. hormones in a blood stream.

The greater the metabolic rate of an organism, the greater are the demands on its mass transport system. Rapid movement through the transport system is improved by an organ which has a pumping mechanism. The heart is an excellent example of how this is achieved.

Mammals have a **double circulation** system. This means that as blood enters the heart it is pumped to the lungs, exchanges oxygen for carbon dioxide, and returns to the heart where further pumping propels it through the rest of the body. The blood moves through the heart twice during each cardiac cycle. This double circulation has, through evolution, enabled some species to achieve a greater size because essential substances can reach cells efficiently, over longer distances. The extra pumping action acts as a boost so that greater distances can be achieved.

5.2 Heart: structure and function

After studying this section you should be able to:

- *recall the structure, cardiac cycle and electrical stimulation of a mammalian heart*

The mammalian heart

AQA A	M1
AQA B	M3
EDEXCEL	M2
OCR	M3
WJEC	M2
NICCEA	M2

The heart consists of a range of tissues. The most important one is cardiac muscle. The cells have the ability to contract and relax through the complete life of the person, without ever becoming fatigued. Each cardiac muscle cell is **myogenic**. This means it has its own inherent rhythm. Below is a diagram of the heart.

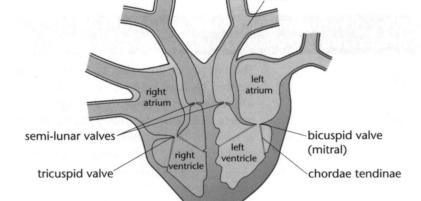

Note that tricuspid and bicuspid valves are known as atrioventricular valves.

Structure

The heart consists of four chambers, **right** and **left atria** above **right** and **left ventricles**. The functions of each part are as follows.

If blood moved in the wrong direction, then transport of important substances would be impeded.

- The **right atrium** links to the **right ventricle** by the **tricuspid valve**. This valve prevents backflow of the blood into the atrium above, when the ventricle contracts.

- The **left atrium** links to the **left ventricle** by the **bicuspid valve (mitral valve)**. This also prevents backflow of the blood into the atrium above.

Check out these diagrams of a valve.

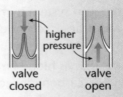

valve   valve
closed   open

You can work out if a valve is open or closed in terms of pressure. Higher pressure above than below a semi-lunar valve closes it. Higher pressure below the semi-lunar valve than above, opens it.

- The **chordae tendonae** attach each ventricle to its **atrioventricular valve**. Contractions of the ventricles have a tendency to force these valves up into the atria. Backflow of blood would be dangerous, so the chordae tendonae hold each valve firmly to prevent this from occurring.

- Semi-lunar (pocket) valves are found in the blood vessels leaving the heart (pulmonary artery and aorta). They only allow exit of blood from the heart through these vessels following ventricular contractions. Contraction of these arteries and relaxation of the ventricles closes each semi-lunar valve.

- Ventricles have thicker muscular walls than atria. When each atrium contracts it only needs to propel the blood a short distance into each ventricle.

- The left ventricle has even thicker muscular walls than the right ventricle. The left ventricle needs a more powerful contraction to propel blood to the systemic circulation (all of the body apart from the lungs). The right ventricle propels blood to the nearby lungs. The contraction does not need to be so powerful.

Cardiac cycle

Blood must continuously be moved around the body, collecting and supplying vital substances to cells as well as removing waste from them. The heart acts as a pump using a combination of systole (contractions) and diastole (relaxation) of the chambers. The cycle takes place in the following sequence.

Stage 1

Ventricular diastole, atrial systole
Both ventricles relax simultaneously. This results in lower pressure in each ventricle compared to each atrium above. The atrioventricular valves open partially. This is followed by the atria contracting which forces blood through the atrioventricular valves. It also closes the valves in the vena cava and pulmonary vein. This prevents backflow of blood.

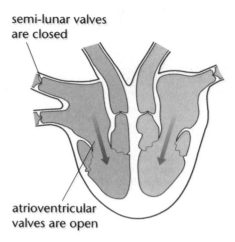

semi-lunar valves are closed

atrioventricular valves are open

Examination questions often test your knowledge of the opening and closing of valves. Always analyse the different pressures given in the question. A greater pressure behind a valve opens it. A greater pressure in front closes it.

Stage 2

Ventricular systole, atrial diastole
Both atria then relax. Both ventricles contract simultaneously. This results in higher pressure in the ventricles compared to the atria above. The difference in pressure closes each atrioventricular valve. This prevents backflow of blood into each atrium. Higher pressure in the ventricles compared to the aorta and pulmonary artery opens the semi-lunar valves and blood is ejected into these arteries. So blood flows through the systemic circulatory system via the aorta and vena cava and through the lungs via the pulmonary vessels.

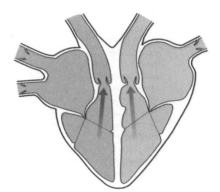

Stage 3

Ventricular diastole, atrial diastole
Immediately following ventricular systole, both ventricles and atria relax for a short time. Higher pressure in the aorta and pulmonary artery than the ventricles closes the semi-lunar valves. This prevents the backflow of blood. Higher pressure in the vena cava and pulmonary vein than the atria results in the refilling of the atria.

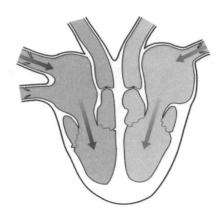

Returning to stage one the cycle begins again. The hormone adrenaline increases the heart rate still further. Even your examinations may increase your heart rate!

The cycle is now complete – **GO BACK TO STAGE 1!**

The whole sequence above is one cardiac cycle or heartbeat and it takes less than one second! The number of heartbeats per minute varies to suit the activity of an organism. Vigorous exercise is accompanied by an increase in heart rate to allow faster collection, supply and removal of substances because of enhanced blood flow. Conversely during sleep, at minimum metabolic rate, heart rate is correspondingly low because of minimum requirements by the cells.

SAN

AVN

Purkinje tissue

> The SAN is the natural pacemaker of the heart.

> All of the Purkinje fibres together are known as the **Bundle of His**.

> This is one of the examiners' favourite ways to test heart-related concepts. Look at the **peak** of the **ventricular contraction**. It coincides with the **trough** in the **ventricular volume**. This is not surprising, because as the ventricle contracts it empties! Use the data of higher pressure in one part and lower in another to explain:
>
> (a) movement of blood from one area to another
>
> (b) the closing of valves.

How is the heart rate controlled?

It has already been stated that the cardiac muscle cells have their own inherent rhythm. Even an individual cardiac muscle cell will contract and relax on a microscope slide under suitable conditions. An orchestra would not be able to play music in a coordinated way without a conductor. The cardiac muscle cells must be similarly coordinated, by electrical stimulation from the brain.

- The heart control centre is in the medulla oblongata.
- The sympathetic nerve stimulates an increase in heart rate.
- The vagus nerve stimulates a decrease in heart rate.
- These nerves link to a structure in the wall of the right atrium, the sinoatrial node (SAN).
- A wave of electrical excitation moves across both atria.
- They respond by contracting (the right one slightly before the left).
- The wave of electrical activity reaches the atrioventricular node (AVN) which conducts the electrical activity through the Purkinje fibres.
- These Purkinje fibres pass through the septum of the heart deep into the walls of the left and right ventricles.
- The ventricle walls begin to contract from the apex (base) upwards.
- This ensures that blood is ejected efficiently from the ventricles.

Graphs to show the changes in pressure and volume during the cardiac cycle

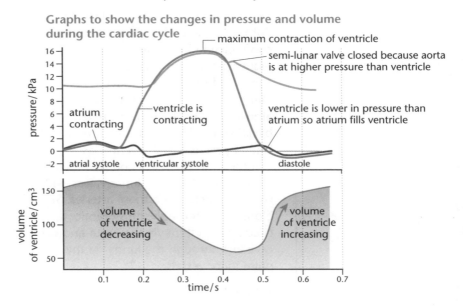

Progress check

The medulla oblongata can increase the heart rate. The statements below include all of the events which take place, but in the wrong order. Write them out in the correct sequence.

A this ensures that blood is ejected efficiently from the ventricles

B the wave of electrical activity reaches the **atrioventricular node (AVN)** which conducts the electrical activity through the **Purkinje fibres**

C a wave of electrical excitation moves across both atria

D the sympathetic nerve conducts electrical impulses

E electrical impulses are received at the **sinoatrial node (SAN)**

F as a result the atria contract

G the ventricle walls begin to contract from the apex (base) upwards

D E C F B G A

5.3 Blood vessels

After studying this section you should be able to:

- *describe the structure and functions of arteries, veins and capillaries*
- *understand the importance of valves in the return of blood to the heart*

LEARNING SUMMARY

Arteries, veins and capillaries

AQA A	M1
AQA B	M3
EDEXCEL	M2
OCR	M3
WJEC	M2
NICCEA	M2

The blood is transported to the tissues via the vessels. The main propulsion is by the ventricular contractions. Blood leaves the heart via arteries, reaches the tissues via the capillaries, then returns to the heart by the veins. Each blood vessel has a space through which the blood passes; this is the lumen. The structure of the vessels is shown below.

Artery

Note that the pressure in the **arteries** is highest because:

(a) they are closest to the ventricles
(b) they contract forcefully themselves.

Capillaries are the next highest in pressure, the main factor being their resistance to blood flow.

Finally, the pressure of **veins** is the lowest because:

(a) they are furthest from the ventricles
(b) they have a low amount of muscle.

If given blood pressures of vessels, be ready to predict the correct direction of blood flow.

- It has a thick tunica externa which is an outer covering of tough collagen fibres.
- It has a tunica media which is a middle layer of smooth muscle and elastic fibres.
- It has a lining of squamous endothelium (very thin cells).
- It can contract using its thick muscular layer.

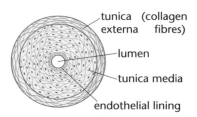

tunica externa (collagen fibres)
lumen
tunica media
endothelial lining

Capillary

- It is a very thin blood vessel, the endothelium is just one cell thick.
- Substances can exchange easily.
- It has such a high resistance to blood flow that blood is slowed down. This gives more time for efficient exchange of chemicals at the tissues.

endothelium
lumen

Vein

- It has a thin tunica externa which is an outer covering of tough collagen fibres.
- It has a very thin tunica media which is a middle layer of smooth muscle and elastic fibres.
- It has a lining of squamous endothelium (very thin cells).
- It is lined with semi-lunar valves which prevent the backflow of blood.

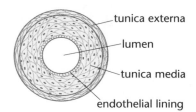

tunica externa
lumen
tunica media
endothelial lining

How do the veins return the blood to the heart?

direction of blood flow

semi-lunar valve

Veins have a thin tunica media, so only mild contractions are possible. They return blood in an unexpected way. Every time the organism moves physically, blood is squeezed between skeletal muscles and forced along the vein.

> It must travel towards the heart because of the direction of the semi-lunar valves. Any attempt at backflow and the semi-lunar valves shut tightly!

KEY POINT

Capillary network

Every living cell needs to be close to a capillary. The arteries transport blood from the heart but before entry into the capillaries it needs to pass through an **arteriole**. The arteriole is a ring of muscle known as a **pre-capillary sphincter**. When this is contracted the constriction shuts off blood flow to the capillaries, but when dilated blood passes through. Some capillary networks have a shunt vessel. When the arteriole is constricted blood is diverted along the shunt vessel so the capillary network is by-passed. After the capillary network has permeated through an organ the capillaries link into a **venule** which joins a **vein**.

In the skin the superficial capillaries have the arteriole/shunt vessel/venule arrangement as shown opposite. When the arteriole is dilated (**vasodilation**) more heat can be lost from the skin. When the arteriole is constricted (**vasoconstriction**) the blood cannot enter the capillary network so is diverted to the core of the body. Less heat is lost from the skin.

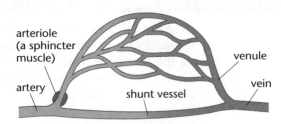

arteriole (a sphincter muscle)

artery

shunt vessel

venule

vein

Blood

The blood consists of a fluid in which many solutes are dissolved and blood cells are suspended. It is constantly circulated around the body. Additionally it has roles in combating infections (see page 135). The shape of a red blood cell is biconcave. This allows a greater amount of oxygen to be transported, because of the greater surface area to volume ratio. It important that the blood has enough red blood cells and that each red blood cell contains enough haemoglobin to transport the oxygen efficiently.

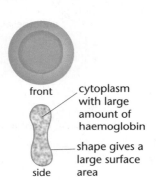

front

cytoplasm with large amount of haemoglobin

shape gives a large surface area

side

The human body can adapt well to different environmental conditions, e.g. if a person who lives close to sea level moves to a much higher altitude, like Mexico City, they experience much thinner air, and take in correspondingly less oxygen. The breathing rate of the person would, initially, be more rapid than usual. After several months the bone marrow makes more red blood cells, so oxygen transport is much improved even though the air is less dense than in the person's normal environment. Athletes use this technique to prepare for important events. Even when they return to compete at lower altitude oxygen transport is so improved that they usually compete very well and often exceed their previous best.

Progress check

The diagram shows the structure of a blood vessel.

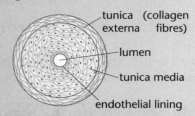

tunica externa (collagen fibres)

lumen

tunica media

endothelial lining

(a) (i) Which type of vessel, artery, vein or capillary is shown? Give a reason for your choice.

(ii) What is the function of the tunica media?

(b) The pressure values 30 kPa, 10 kPa and 5 kPa correspond to the different types of vessel. Give the correct value for each vessel so that blood flows around the body.

(a) (i) artery; the vessel has a thick tunica externa
(ii) contracts to help transport blood.
(b) artery, 30 kPa, capillary, 10 kPa and vein, 5 kPa.

5.4 The transport of substances in the blood

After studying this section you should be able to:

- *describe the transport of oxygen in the blood and explain how oxygen is released at the tissues*
- *describe the transport of carbon dioxide*

How is oxygen transported?

AQA B	M3
EDEXCEL	M2
OCR	M3
WJEC	M2
NICCEA	M2

Oxygen is absorbed in the lungs from fresh air which has been breathed in. Red blood cells (erythrocytes) contain the protein haemoglobin which has an affinity (attraction) for oxygen. This means that oxygen readily binds with the haemoglobin. Even when oxygen is in short supply oxygen will effectively bind with the haemoglobin and the red blood cells will carry the oxyhaemoglobin.

The red blood cells have no nucleus, which increases the surface area to volume ratio, which increases the amount of oxygen taken up. Efficient transport is important but crucial is the ability to give up the oxygen to the tissues which need it. The Bohr effect explains the way that oxygen moves from the red blood cells to the tissues. It is shown by the graph below.

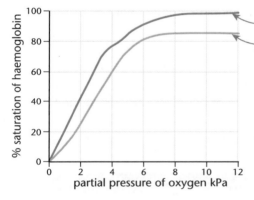

Haemoglobin takes up oxygen in the lungs.

At the tissues, the graph has been pushed to the right and down. Some oxygen leaves the haemoglobin and goes to the tissues.

Features of the Bohr effect

- The graph is known as the oxygen dissociation curve and its shape is sigmoid ('S' shaped).

> The term **oxygen tension** can be used instead of partial pressure.

- Even in a low partial pressure of oxygen, the percentage saturation of haemoglobin is very high as shown by the steep incline on the graph at low partial pressures.

- Haemoglobin holds the oxygen strongly as the blood is transported.

- Once they reach the tissues, the red blood cells encounter carbon dioxide which causes the dissociation curve to move downwards and to the right.

> The answers to Bohr effect questions revolve around:
> (a) the uptake of oxygen by haemoglobin even at low partial pressures
> (b) the 'offloading' of oxygen by carbon dioxide at the tissues.

- The net effect is that some oxygen is released from the haemoglobin and supplied to the tissues.

- The greater the amount of carbon dioxide at the tissues, the more the dissociation curve is moved to the right, and the more oxygen is 'off-loaded' to the tissues.

Fetal haemoglobin has a greater affinity for oxygen than adult haemoglobin. This allows the fetus to take oxygen from the mother's haemoglobin. There is a substance with a greater affinity for oxygen than haemoglobin. It is myoglobin. Sea lions have high quantities in their muscles. The myoglobin acts as an oxygen store so the animals can remain under water for a long period of time.

Lymphatic system

AQA A	M1
AQA B	M3
EDEXCEL	M2
OCR	M3
WJEC	M2
NICCEA	M2

There is a network of vessels other than the blood system. They are the **lymphatic vessels**. They collect surplus tissue fluid, similar to blood plasma.

The lymph vessels have valves to ensure transport is in one direction. Along some parts of the lymphatic system are lymph nodes. These are swellings lined with white blood cells (macrophages and lymphocyte cells, see page 135).

Transport of carbon dioxide

AQA A	M1
AQA B	M3
EDEXCEL	M2
OCR	M3
WJEC	M2
NICCEA	M2

This is done with the help of the red blood cells as follows:

- carbon dioxide diffuses into red blood cells from the tissues
- the carbon dioxide reacts with water to produce carbonic acid, this reaction being catalysed by the enzyme **carbonic anhydrase** in the cell (*a very fast reaction!*).

$$\text{carbonic anhydrase}$$
$$H_2O + CO_2 = H_2CO_3$$
water   carbon   carbonic
     dioxide   acid

- the carbonic acid **ionises** into H^+ and HCO_3^-

$$H_2CO_3 = H^+ + HCO_3^-$$

- haemoglobin combines with H^+ ions forming **haemoglobinic acid** which is very weak

$$H^+ + Hb = HHb$$

- HCO_3^- ions diffuse into the blood plasma to be transported to the lungs
- Cl^- ions diffuse into the red blood cell from the plasma; this counteracts the build up of positive charge from the H^+ ions. This is known as the **chloride shift**.

The whole process is reversed once the blood reaches the lungs.

> Take great care when answering questions about carbon dioxide transport. You will need to give details about the role of the red blood cell in carbon dioxide transport. Just stating that the red cell transports carbon dioxide is wrong! The process is much more complex than that and the HCO_3^- ions diffuse out of the red cells into the plasma.

Plasma

AQA A	M1
AQA B	M3
EDEXCEL	M2
OCR	M3
WJEC	M2
NICCEA	M2

This is the fluid in which all of the blood contents are transported. Listed below are some substances transported in the plasma:

- **water** – dissolves substances such as glucose for transport, stores dissolved prothrombin and fibrinogen which may be used later in clotting
- **proteins** – some are used to buffer the pH of the blood
- **glucose** – on its way to releasing energy in respiration
- **lipids** – on their way to releasing energy in respiration
- **amino acids** – on their way to cells to help assemble proteins or release energy in respiration
- **salts** – contribute to the water potential of blood, so that cells are not dehydrated by osmosis
- **hormones** – chemical messenger-molecules on their way to a target organ
- **antigens** – recognition proteins preventing white blood cells from destroying the person's own blood
- **antibodies** – made by lymphocytes to destroy antigens
- **urea** – made in the liver from excess amino acids, extracted by the kidneys.

Blood has a major role in the defence against disease (see the immune system page 134).

> Water has many important functions in the body, including being transported to the sweat glands to cool the body down.

> Note that the list outlines just some of the functions of plasma-transported substances. There are many more!

5.5 The transport of substances in a plant

After studying this section you should be able to:

LEARNING SUMMARY

- *recall the structure of a root and understand how water and ions are absorbed*
- *recall the structure of xylem and phloem and explain the processes by which they transport essential chemicals*

Root structure and functions

AQA B	M3
EDEXCEL	M2
OCR	M1, M3
WJEC	M2
NICCEA	M2

Note that the root hairs also absorb oxygen from the air to aid aerobic respiration. The high surface area to volume ratio certainly helps!

Remember that water moves from a higher water potential to a more negative water potential.

Remember that active transport needs energy, so mitochondria will be close to the carrier molecules on the membranes.

The roots of a green plant need to exchange substances with the soil environment. The piliferous zone just behind a root tip has many root hairs which have a high surface area to volume ratio.

- Root hairs are used for absorption of water and mineral ions and the excretion of carbon dioxide.
- They have a cell membrane with a high surface area to volume ratio to efficiently absorb water, mineral ions and oxygen and excrete carbon dioxide.
- They project out into the soil particles which are surrounded by soil water at high water potential compared to the low water potential of the contents of the root hairs.
- They have a cell membrane which is partially permeable to allow water absorption by osmosis (see page 59).
- As they absorb more water by osmosis, a cell sap becomes more dilute compared to neighbouring cells. Water therefore moves to these adjacent cells which become more diluted themselves, so osmosis continues across the cortex.
- They have carrier proteins in the cell membranes to allow mineral ions to be absorbed by active transport.

Passage of water into the vascular system

Note the different theories for water transport across the width of the root.

Once absorbed by osmosis, water needs to pass to the xylem vessels in order to move up the plant. First it must move across the cortex of the root and through the endodermis before entering the xylem. The mechanism of passage is not known but there are three theories:

- **apoplast** route, where the water is considered to pass on the outside of the cells
- **symplast** route, where the water is considered to pass via the cytoplasm of the cells via **plasmodesmata** (cytoplasmic strands connecting one cell to another)
- **vacuolar** route, where the water is considered to pass through the tonoplast then through the sap vacuole of each cell.

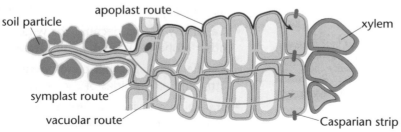

movement of water across the cortex

Casparian strip

Water moves across the cortex and needs to pass through the endodermal cells before entering the xylem vessels of the vascular system. Each cell of the endodermis has a waterproof band around it, just like a ribbon around a box. This means that water must pass through the cell in some way, rather than around the outside. If water moves by the apoplast route up to this point, then it must now move into the symplast or vacuolar pathways.

water must pass through middle of cell

Casparian strip (waterproof band)

Casparian strip of the endodermal cells

Once the water has passed through the endodermis and navigated the pericycle then it must pass into the xylem for upward movement to the leaves and to the tissues.

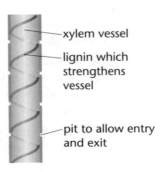

xylem vessel

lignin which strengthens vessel

pit to allow entry and exit

How does water move up the vascular system to the leaves?

Water moves into the xylem vessels in the vascular system in the centre of the root; it enters via bordered pits. The xylem is internally lined with lignin. This substance is waterproof and it also gives great strength to the xylem vessels, which are tube shaped. Much of the strength of a plant comes from cells toughened by lignin. A Giant Redwood tree is many metres high but water is still able to reach all the cells. Water moves up the xylem for the following reasons.

- Root pressure gives an initial upward force to water in the xylem vessels. This can be shown by cutting off a shoot near soil level. Some sap will pour vertically out of the xylem of the remaining exposed xylem.
- Water moves up the xylem by capillarity which is the upward movement of a fluid in a narrow bore tube – xylem has very narrow vessels.
- Capillarity occurs because the water molecules have an attraction for each other (cohesion) so when one water molecule moves others move with it.
- Capillarity has another component – the fact that the water molecules are attracted to the sides of the vessels pulls the water upwards (adhesion).
- Transpiration causes a very negative water potential in the mesophyll of the leaves. Water in the xylem is of higher water potential and so moves up the xylem.

Remember that the xylem is part of the mass flow system ensuring that all cells receive their requirements.

The factors in the list are known as the cohesion-tension theory and explain how water moves up the xylem.

Xylem vessels die at the end of their maturation phase. The lignin produced inside the cells finally results in death. The young xylem cells end to end, finally produce a long tube-like structure (vessel) through which water passes. Xylem can still transport water after the death of the plant.

Mineral ions are also transported in the xylem.

Translocation

This is an active process by which sugars and amino acids are transported through the phloem. Sugar is produced in the photosynthetic tissues and must be exported from these sources to areas of need, i.e. usually areas which have large energy requirements. These areas are called sinks, e.g. terminal buds and roots.

> Roots cannot photosynthesise so they need carbohydrates to be supplied by other parts of the plant such as the leaf or storage organs.

KEY POINT

The sugars are transported in the phloem which consists of two types of cell, the sieve tube and companion cell. Unlike the xylem the cells of the phloem are living.

Structure of the phloem tissue

The sieve tube has no nucleus so that essential proteins for life are made by the companion cell which does possess a nucleus. The companion cell maintains services to the sieve tube.

- Each sieve tube links to the next via a sieve plate which is perforated with pores.
- The sieve tube has cytoplasm and a few small mitochondria.
- Sugars are thought to pass through the sieve tubes by cytoplasmic streaming.
- The sieve tubes have no nucleus but are alive because of cytoplasmic connections (plasmodesmata) with the companion cell.
- Each companion cell has a nucleus and mitochondria.

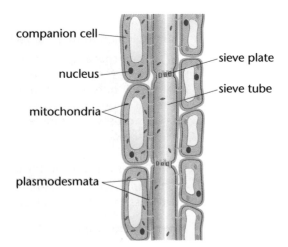

Radioactive labelling

This technique has been used to investigate the mechanism of translocation. It involves the use of a substance such as radioactive CO_2. The radioactive isotope, ^{14}C is used to make $^{14}CO_2$. A leaf is allowed to photosynthesise in the presence of $^{14}CO_2$ and makes the radioactive sugar ($^{14}C_6H_{12}O_6$). The route of the radioactive sugar can be traced using a Geiger-Mueller counter. The greater the number of radioactive disintegrations per unit time, the greater is the concentration of the sugar in that part of the plant

5.6 Water loss in a plant

After studying this section you should be able to:

- understand how guard cells open and close
- understand the process of transpiration
- know how to use a potometer to measure the rate of transpiration
- recall the adaptations of xerophytes

LEARNING SUMMARY

How is water lost from a leaf?

AQA A	M2
AQA B	M3
EDEXCEL	M2
OCR	M3
WJEC	M2
NICCEA	M2

Water moves up the xylem and into the mesophyll of a leaf. The process by which water is lost from any region of a plant is transpiration. Water can be lost from areas such as a stem, but most water is lost by evaporation through the stomata. Each stoma is a pore which can be open or closed and is bordered at either side by a guard cell. The diagrams show an open stoma and a closed stoma.

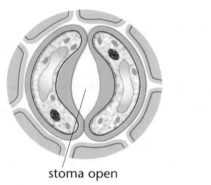

stoma open

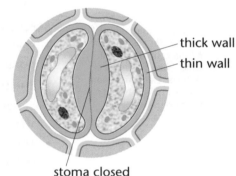

thick wall
thin wall

stoma closed

Transpiration from a leaf takes place as follows:

- the air spaces in the mesophyll become saturated with water vapour (higher water potential)
- the air outside the leaf may be of lower humidity (more negative water potential)
- this causes water molecules to diffuse from the mesophyll of the leaf to the outside.

KEY POINT

Some water can escape through the cell junctions and membranes. This is known as cuticular transpiration. In the dark all stomata are closed. Even so, there is still water loss by cuticular transpiration.

How do the guard cells open and close?

In the presence of light:

- K$^+$ ions are actively transported into the guard cells from adjacent cells
- malate is produced from starch
- K$^+$ ions and malate accumulate in the guard cells
- this causes an influx of water molecules
- the cell wall of each guard cell is thin in one part and thick in another
- the increase in hydrostatic pressure leads to the opening of the stomata.

Closing of the stomata is the reverse of this process. Under different conditions the stomata can be partially open. The rate of transpiration can increase in warm, dry conditions or decrease at the opposite extreme.

Measuring the rate of transpiration

This is done indirectly by using a potometer. This instrument works by the following principle; for every molecule of water lost by transpiration, one is taken up by the shoot.

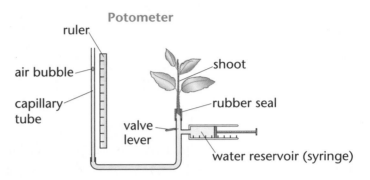

Potometer

The potometer is used as follows:

- a shoot is cut and the end is quickly put in water to prevent an air lock in the xylem
- the potometer is filled under water so that the capillary tube is full
- all air bubbles are removed from the water
- the shoot is put into the rubber seal
- the valve is changed to allow water uptake
- the amount of water taken up by the shoot per unit time is measured
- the shoot can be tested under various conditions.

Note that a living shoot may be photosynthesising whilst attached to the instrument. Only a minute amount of water would be used in this process. The instrument gives an accurate measure of transpiration.

Xerophytes

AQA A	M2
AQA B	M3
EDEXCEL	M2
OCR	M3
WJEC	M2
NICCEA	M2

These are plants which have special adaptations to survive in drying, environmental conditions where many plants would become desiccated and die. The plants survive well because of a combination of the following features:

- thick cuticle to reduce evaporation
- reduced number of stomata
- smaller and fewer leaves to reduce surface area
- hairs on plant to reduce air turbulence
- protected stomata to prevent wind access
- aerodynamic shape to prevent full force of wind
- deep root network to absorb maximum water
- some store water in modified structures, e.g. the stem of a cactus.

In an exam you may be given a photomicrograph of a xerophytic plant which you have not seen before. Look for *some* of the features covered in the bullet points opposite.

A cactus is an excellent example of a xerophyte. It makes excellent use of what little water there is available, and holds on to what it does manage to absorb really well. Its cuticle and epidermis are so thick that metabolic water released from the cells during night time respiration is retained for photosynthesis during the day. Nothing is wasted!

Progress check

Water is absorbed into a plant by the root hairs.

(a) The water potential of the root hair cells are more negative than in the soil water. Is this statement true or false?

(b) Describe:
 (i) the apoplast route across the cortex
 (ii) the symplast route across the cortex.

(a) true

(b) (i) water is considered to pass on the outside of the cell membrane
 (ii) water passes through the cytoplasm of the cells through plasmodesmata.

Sample question and model answer

(a) The graph below shows the oxygen dissociation curve for human haemoglobin.

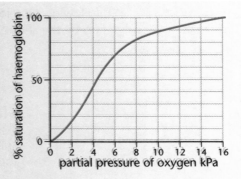

Note that haemoglobin is able to pick up a lot of oxygen, even at low partial pressure.

Use the information in the graph to help you answer the following questions.

(i) What is the advantage of haemoglobin as a respiratory pigment when oxygen in the air is at the low partial pressure 6 kPa? [1]

Even at a low partial pressure a lot of oxygen (70%) is taken up by the haemoglobin of a red blood cell.

The fact that haemoglobin is able to carry oxygen is important. However, it is just as important that the oxygen is offloaded at tissues needing it. This is only possible because carbon dioxide is found at the tissues.

(ii) Explain the effect on the oxygen dissociation curve of a high partial pressure of carbon dioxide at a muscle. [2]

The curve is moved to the right and down so that oxygen is released.

(iii) Fetal haemoglobin has a greater affinity for oxygen than maternal haemoglobin. Draw a curve on the graph to show the oxygen dissociation curve for fetal haemoglobin. [1]

See graph opposite.

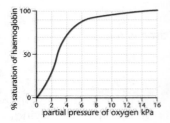

(b) The diagram shows **one** stage in the cardiac cycle.

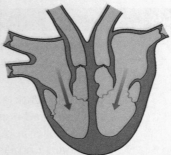

Always look for the valves. If the heart valve is open then the chamber behind it is contracting.

(i) Which stage of the cardiac cycle is shown in the diagram? Give **two** reasons for your answer. [3]

atrial systole
the atrioventricular valves are open/blood flows through the atrioventricular valves,
semi-lunar valves are closed.

(ii) Write an X in one chamber to show the position of the atrioventricular node (AVN). [1]

(iii) How does the AVN stimulate the contraction of the ventricles? [1]

Passes electrical impulses to Purkinje tissue/Bundle of His.

Practice examination questions

1 The diagram shows a capillary bed in the upper part of the skin. The arteriole is constricted.

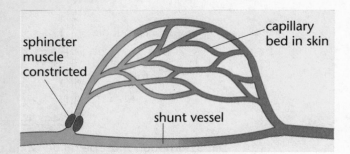

Use the information in the diagram and your own knowledge to answer the questions below.

(a) As a result of arteriole constriction, to where would the blood flow? [1]

(b) Explain how this would help maintain the body temperature. [4]

2 The table shows data about a person's heart before and after a training programme.

	Before training	After training
heart stroke rate	90 ml	120 ml
heart rate at rest	75 bpm	60 bpm
maximum heart rate	170 bpm	190 bpm

(a) Over a five minute period at rest before training, the cardiac output of the person was 33.75 litres.

How much blood would leave the heart, during the same time, whilst the person was at rest, after training? [2]

(b) After training, the maximum heart rate increased by 20 bpm. Explain the advantage of this increase to an athlete. [4]

(c) After training there are other changes in the body.

Explain:

(i) **two** changes which would improve the efficiency of the respiratory system. [2]

(ii) the effect of training on the muscles. [2]

3 Smoking causes respiratory disease. Outline the course and symptoms of each disease:

(a) lung cancer

(b) bronchitis

(c) emphysema. [6]

Practice examination questions (continued)

4   The diagram shows a freshly cut, leafy shoot attached to a potometer. This was used to measure the amount of water taken up by the shoot under different conditions.

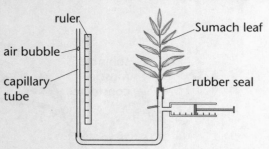

(a) What assumption must be made when using this apparatus to measure the rate of transpiration?   [1]

(b) An air-lock can occur in the shoot which prevents water uptake.

   (i)   In which plant tissue could an air-lock occur?   [1]
   (ii) Describe the practical details by which a student could make sure that there was no air-lock in the shoot.   [2]

(c) The radius of the capillary tube of the potometer was 1 mm. When a Sumach leaf was measured the air bubble moved 32 mm in one minute. Calculate the volume of water in mm³ which would be taken up by the leaf in one hour under the same environmental conditions.   [3]

5   *Agave americana* is a xerophytic plant which grows in the deserts of Mexico.

Agave americana

Suggest **three** ways in which the plant is adapted to survive periods of very low rainfall.   [3]

6   The diagram shows nerves linking the medulla oblongata with the heart.

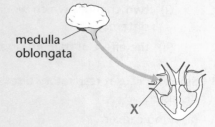

(a) Name part X.   [1]

(b) What effect do the following have on the heart:

   (i)   vagus nerve
   (iii) sympathetic nerve
   (iv) adrenaline?   [3]

Practice examination questions (continued)

7 The diagram shows an aphid feeding on a plant. The sharp stylet is inserted into the phloem tissue which supplies the aphid with sucrose, plus organic and inorganic ions.

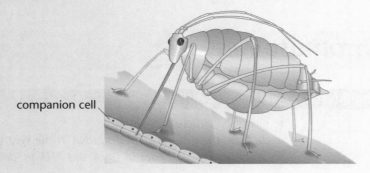

companion cell

(a)

(i) Name the phloem cell X from which the aphid obtains sucrose. [1]

(ii) Cell X does not have a nucleus or ribosomes, but still contains enzymes. Explain how this is possible. [3]

(b)

(i) Feeding aphids obtain the contents of the phloem without any sucking action being necessary. What does this indicate about the transport of substances through the phloem? [3]

(ii) Scientists investigated phloem contents by anaesthetising feeding aphids, then cutting their bodies from their stylets. Phloem contents oozed from the cut end of each stylet. The phloem contents were tested using iodine and heating with Benedict's solution.

	Tested with iodine	*Heated with Benedict's solution*
Contents of phloem	brown colour	brick-red colour

Referring to the results of the tests, explain what the scientists found out about the phloem contents using this method. [4]

(iii) Hot-wax ringing is a technique where hot wax is poured around a stem. This technique was used with the aphid method described in (ii). Radioactive carbon dioxide was supplied to one leaf so that a radioactive carbohydrate was made.

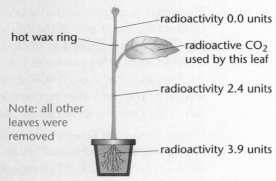

radioactivity 0.0 units

hot wax ring

radioactive CO_2 used by this leaf

radioactivity 2.4 units

Note: all other leaves were removed

radioactivity 3.9 units

Explain the effect of hot-wax ringing on the phloem tissue. [3]

Chapter 6
The genetic code

The following topics are covered in this chapter:

- Chromosomes
- Cell division
- Gene technology
- Genetically modified organisms

6.1 Chromosomes

After studying this section you should be able to:

- *recall the structure of DNA*
- *describe the roles of DNA and RNA in the synthesis of protein*
- *use organic base codes of DNA and RNA to identify amino acid sequences*

LEARNING SUMMARY

Chromosome structure and function

AQA A	M2
AQA B	M2
EDEXCEL	M1
OCR	M1
WJEC	M1
NICCEA	M1

Each chromosome in a nucleus consists of a series of genes. A gene is a section of **DNA** and each controls the production of a protein important to the life of an organism.

Deoxyribonucleic acid (DNA)

You need to be aware that many nucleotides join together to form the polymer, DNA.

Deoxyribonucleic acid (DNA) is made up of a number of **nucleotides** joined together in a double helix shape.

Why does the DNA of one organism differ from the DNA of another?

The answer lies in the structure of their nucleotides. Look at the structure of one nucleotide – **monomer**.

Each strand of DNA is said to be complementary to the other. **Examination tip:** be ready to identify one strand when given the matching complementary strand.

The organic base of each nucleotide can be any one of **adenine**, **thymine**, **cytosine** or **guanine**. Nucleotides join together at their bases by hydrogen bonds. Adenine bonds with thymine and cytosine with guanine.

Phosphate and pentose sugar units link to form the backbone of the DNA. The twisting pattern formed as nucleotides bond to each other produces the double helix shape of DNA. Repeated linking of the monomer nucleotides forms the polymer structure of DNA.

Differences in the DNA of organisms such as humans and houseflies lie in the **different sequences** of the organic bases. Each sequence of bases is a code to make a protein, usually vital to the life of an organism.

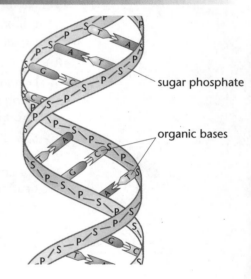

sugar phosphate

organic bases

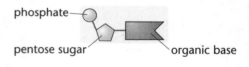

A single nucleotide

phosphate

pentose sugar

organic base

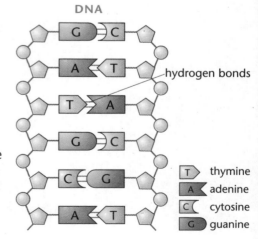

DNA

hydrogen bonds

T	thymine
A	adenine
C	cytosine
G	guanine

How does a cell make protein?

This is called **protein synthesis**. It begins by linking amino acids in a chain sequence to form a **polypeptide**. Later a number of polypeptides can bond together to form an entire protein. The following diagrams show protein synthesis.

1 In the nucleus **RNA polymerase** links to a start code along a DNA strand.

2 RNA polymerase moves along the DNA. For every organic base it meets along the DNA a complementary base is linked to form mRNA (**messenger RNA**).

There is no thymine in mRNA. Instead there is another base, uracil.

Transcription

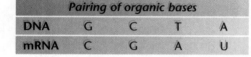
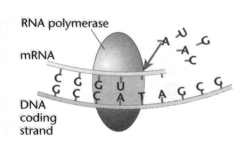

Pairing of organic bases				
DNA	G	C	T	A
mRNA	C	G	A	U

3 RNA polymerase links to a finish code along the DNA and finally the mRNA **moves to a ribosome**. The DNA stays in the nucleus for the next time it is needed.

4 Every three bases along the mRNA make up one **codon** which codes for a specific amino acid. Three complementary bases form an **anti-codon** attached to one end of tRNA (**transfer RNA**). At the other end of the RNA is a specific amino acid.

Note the link between each pair of amino acids along a polypeptide – the peptide link.

5 All along the mRNA the tRNA 'partner' molecules enable each amino acid to bond to the next. A chain of amino acids (**polypeptide**) is made, ready for release into the cell.

Translation

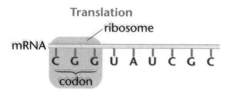
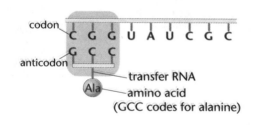

amino acid (GCC codes for alanine)

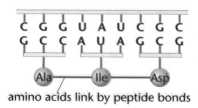

amino acids link by peptide bonds

DNA codes

The table below shows all the triplet sequences of organic bases found along DNA strands and the coding function of each. During the process of protein synthesis triplet codes along the DNA result in specific amino acids being linked in chains known as **polypeptides**. Firstly, each triplet in the table codes for the production of complementary bases along mRNA.

Here is an example of a coding strand of DNA

| DNA | A A A G A G G A C A C T | (coding strand) |
| mRNA | U U U C U C C U G U G A | (messenger RNA) |

Use the key to identify the amino acids in the table opposite.

Amino acid	Abbreviation
alanine	Ala
arginine	Arg
asparagine	Asn
aspartic acid	Asp
cysteine	Cys
glutamine	Gln
glutamic acid	Glu
glycine	Gly
histidine	His
isoleucine	Iso
leucine	Leu
lysine	Lys
methionine	Met
phenylalanine	Phe
proline	Pro
serine	Ser
threonine	Thr
tryptophan	Trp
tyrosine	Tyr
valine	Val

Do not learn all of the triplet codes. Be ready to use the supplied data in the examination. You will be given a key of different codes and functions.

If you are given a table of codes check them carefully. If the bases are from mRNA then there will be uracil in the table.

guanine (G) on DNA codes for cytosine (C) on mRNA

cytosine (C) on DNA codes for guanine (G) on mRNA

thymine (T) on DNA codes for adenine (A) on mRNA

adenine (A) on DNA codes for uracil (U) on mRNA

Special note: There is no thymine found on mRNA. Instead, the organic base uracil is found.

Genetic code functions of DNA

			second organic base							third organic base
		A		G		T		C		
A	A A A	Phe	A G A		A T A	Tyr	A C A	Cys	A
	A A G		A G G	Ser	A T G		A C G		G
	A A T	Leu	A G T		A T T	stop	A C T	stop	T
	A A C		A G C		A T C	stop	A C C	Trp	C
G	G A A		G G A		G T A	His	G C A		A
	G A G	Leu	G G G	Pro	G T G		G C G	Arg	G
	G A T		G G T		G T T	Gln	G C T		T
	G A C		G G C		G T C		G C C		C
T	T A A		T G A		T T A	Asn	T C A	Ser	A
	T A G	Ile	T G G	Thr	T T G		T C G		G
	T A T		T G T		T T T	Lys	T C T	Arg	T
	T A C	Met	T G C		T T C		T C C		C
C	C A A		C G A		CTA	Asp	C C A		A
	C A G	Val	C G G	Ala	CTG		C C G	Gly	G
	C A T		C G T		CTT	Glu	C CT		T
	C A C		C G C		CTC		C C C		C

Each triplet code is non-overlapping. This means that each triplet of three bases is a code, then the next three, and so on along the DNA.

- AAA codes for the amino acid phenylalanine
- GAG codes for the amino acid leucine
- GAC codes for the amino acid leucine
- ACT codes for a stop, the mRNA to be released at the end of transcription.

There are more triplet codes than there are amino acids. This is known as the degenerate code, because an amino acid such as leucine can be coded for by up to six different codes.

Mutation

This is a change in the DNA of an organism. There are different effects on DNA causing different types of mutation. Here is a strand of DNA before mutation.

C T A T C G C A A A T A C G T

Mutation type 1   C T A T C G C A A A T A T G C

This is inversion: The TGC triplet now codes for a different amino acid.

Mutation type 2   C T A T C G C A A A T A C G T C A A

This is addition: The CAA triplet now codes for an extra amino acid.

Mutation type 3   C T A T C G C A A A T A

CGT is missing. This is deletion. One amino acid is missing from the polypeptide.

Mutation type 4

Large sections of DNA can be added, whole chromosomes or sets of chromosomes. Mutation types 1 to 3 are point mutations because they affect one amino acid code.

Addition is a key type of mutation. A regular question is to ask why addition of one base may be more harmful than adding a triplet of new bases. One base may change every amino acid code along the DNA whereas one triplet may add just one amino acid.

Progress check

(a) Name the parts of a nucleotide.

(b) (i) By which bonds do the two strands of DNA link together?

    (ii) How would these bonds be broken in the laboratory to produce single strands of the DNA?

(c) Which organic base is found in DNA but not in RNA?

(d) During protein synthesis:

    (i) which enzyme enables the mRNA to be produced?

    (ii) which process produces mRNA?

    (iii) which process enables amino acids to link together as tRNA 'reads' the mRNA codons?

(a) pentose sugar, phosphate and organic base. The organic base may be thymine, adenine, cytosine, or guanine

(b) (i) hydrogen bonds (ii) heat

(c) thymine (d) (i) RNA polymerase (ii) transcription (iii) translation

6.2 Cell division

After studying this section you should be able to:

- *describe and explain the semi-conservative replication of DNA*
- *understand that DNA must replicate before cell division can begin*
- *recognise each stage of cell division by mitosis*

LEARNING SUMMARY

How do cells prepare for division?

AQA A	M2
AQA B	M2
EDEXCEL	M2
OCR	M1
WJEC	M1
NICCEA	M1

Before cells divide they must first make an exact copy of their DNA by using a supply of organic bases, pentose sugar molecules and phosphates. This is known as the semi-conservative replication of DNA. The diagram (right) shows this taking place.

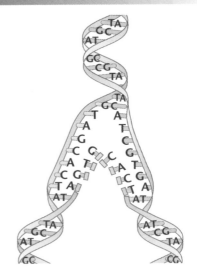

Remember that as the DNA unwinds each single strand is a **complement** to the other. This means that each has the **matching** series of organic bases.

The DNA begins to unwind, exposing its two single strands. Each complementary strand then acts as a template to build its opposite strand. This process results in the production of two identical copies of double stranded DNA.

Semi-conservative replication of DNA

What does semi-conservative mean?

Exam questions are often based on experimental data. Apply your knowledge of principles learned during the course and this will be helpful.

The answer lies in the results of this experiment carried out by researchers.

Bacteria were cultured with a radioactive isotope of nitrogen located in the organic bases of their DNA.

The bacteria were then supplied with non-radioactive bases. They replicated their DNA using these bases. Their population increased.

Each molecule of DNA of the next generation had one radioactive strand and one normal strand.

Semi-conservative replication

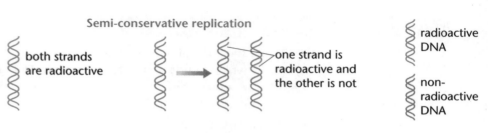

both strands are radioactive → one strand is radioactive and the other is not

radioactive DNA

non-radioactive DNA

Semi-conservative means that as DNA splits into its two single strands, each of the new strands is made of newly acquired bases. The other strand, part of the original DNA, remains.

Cell division

Cells divide for the purposes of growth, repair and reproduction. Not all cells can divide but there are two ways in which division may occur, i.e. mitosis and meiosis.

Just before either mitosis or meiosis begin, interphase takes place. This is when the DNA of the chromosomes replicates (see page 85). The sequence of diagrams below shows a cell dividing into two daughter cells by mitosis.

Remember that DNA replication takes place before cell division in **interphase** (see page 85). This is not an integral phase of mitosis or meiosis.

1 Prophase

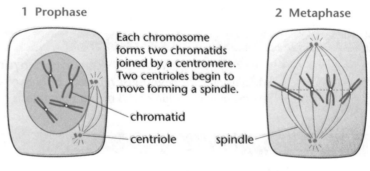

Each chromosome forms two chromatids joined by a centromere. Two centrioles begin to move forming a spindle.

chromatid

centriole

2 Metaphase

The chromatids, still joined by a centromere move to middle of cell. Each of the two chromatids has identical DNA to the other.

spindle

Be ready to analyse photomicrographs of all phases of mitosis. If you can spot 10 pairs of chromosomes at the end of telophase, then this is the original diploid number of the parent cell.

3 Anaphase

The spindle fibres join to the centromeres. The spindle fibres shorten and the centromeres split. The separated chromatids are now chromosomes.

4 Telophase

Identical chromosomes move to each pole. The nuclear membrane re-forms. The cell membrane narrows at the middle and two daughter cells are formed.

The table below shows differences between mitosis and meiosis.

	Mitosis	*Meiosis*
How many daughter cells are produced?	2	4
Are the daughter cells identical or different to the parent cell?	identical (clones)	different
Are the chromosomes of daughter cells single or in pairs?	in pairs (diploid)	single (haploid)

6.3 Gene technology

LEARNING SUMMARY

After studying this section you should be able to:

- define genetic engineering
- describe and explain the roles of key enzymes in genetic engineering
- understand the process of electrophoresis and recall its applications

Manipulating DNA

AQA A	M2
AQA B	M2
OCR	M1
NICCEA	M1

Scientists have developed methods of manipulating DNA. It can be transferred from one organism to another. Organisms which receive the DNA then have the ability to produce a new protein. This is one example of genetic engineering.

> **KEY POINT**
>
> Changes in the DNA of an organism by a range of methods use the knowledge and skills acquired during research into gene technology. The examples which follow show different ways of utilising gene technology.

Gene transfer

AQA A	M2
AQA B	M2
OCR	M1
NICCEA	M1

The gene which produces human insulin was transferred from a human cell to a bacterium. The new microbe is known as a transgenic bacterium. The process which follows shows a similar technique.

Restriction endonucleases are produced by some bacteria as a defence mechanism. They cut up the DNA of invading viruses. This can be exploited during gene transfer.

1 An enzyme known as restriction endonuclease cuts the DNA and the gene was removed. Each time a cut was made the two ends produced were known as 'sticky ends'.

2 Circles of DNA called plasmids are found in bacteria.

Note that **both** the donor DNA and recipient plasmid DNA are cut with the same enzyme. This allows the new gene to be a matching fit.

3 A plasmid was taken from a bacterium and cut with the same restriction endonuclease.

4 The human gene was inserted into the plasmid. It was made to fix into the open plasmid by another enzyme known as ligase.

Many exam candidates fail to state that the plasmids are cloned inside the bacterium.

5 The plasmid replicated inside the bacterium.

The bacteria themselves are also cloned. There may be two marks in a question for each cloning point!

6 Large numbers of the new bacteria were produced. Each was able to secrete perfect human insulin, helping diabetics all over the world.

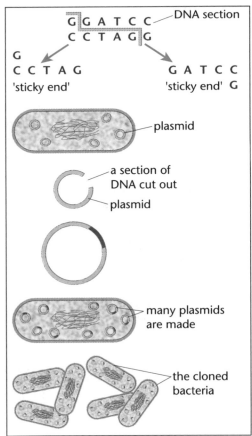

Transgenic bacteria are cultured in huge industrial fermenters, like the one on the next page. They secrete their products which can be collected. Many microbes are exploited in this way. The enzymes found in biological washing powders are produced in this way. Where microbes are exploited for human use, this is known as biotechnology.

Modern industrial fermenters

New transgenic organisms are continually being developed. Many hit the headlines in the media. Be aware that the examiners may use these high-profile organisms in questions. Do not become disorientated – the principle is always the same.

- There are several different types of fermenter used to grow microorganisms on a large scale. They all have the common purpose of producing food, or chemicals such as antibiotics, hormones, or enzymes. The fermenter (right) shows a typical design.

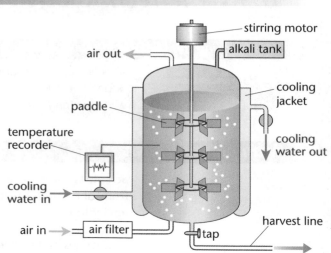

Conditions inside fermenters should be suitable for the optimal metabolism and rapid reproduction of the microorganisms. Products should be harvested without contamination. Note the conditions which need to be controlled.

- Fermenters are sterilised using steam before adding nutrients and the microorganisms used during the process. Conditions are aseptic.
- Nutrients which are specifically suited to the needs of the microorganisms are supplied.
- Air is supplied if the process is aerobic. This must be filtered to avoid contamination from other microorganisms.
- Temperature must be regulated to keep the microorganisms' enzymes within a suitable range. An active 'cooling jacket' and heater, both controlled via a thermostat, enable this to be achieved.
- pH must remain close to the optimum. Often the development of low pH during fermentation would result in the process slowing down or stopping. The addition of alkaline substances allows the process to continue and maximises yield.
- Paddle wheel mixing or 'bubble agitation' make sure that the microorganisms meet the required concentrations of nutrients and oxygen.

Questions about fermentation usually explore your understanding of how the optimal (best) conditions can be achieved. Also graphs may be given. You may be asked about the best time for harvesting the product. This will be at the earliest stage of the graph becoming level.

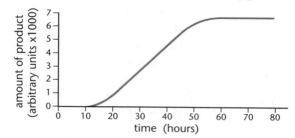

Electrophoresis

Restriction endonucleases can be used to cut up an organism's DNA (see page 87).

- DNA sections are put into a well in a slab of agar gel.
- The gel and DNA are covered with buffer solution which conducts electricity.
- Electrodes apply an electrical field.
- Phosphate groups on DNA are negatively charged causing DNA to move towards the anode.
- Smaller pieces of DNA move more quickly down the agar track, larger ones move more slowly, leading to the formation of bands.

a view looking down on the agar slab

well — Cathode (–ve)
gel
band
track
A   B   C — Anode (+ve)

Can you spot which two samples were from the same person?

Genetic fingerprinting is applied to organisms other than humans. Illegal egg collectors have been successfully prosecuted when egg DNA has been compared against the DNA profile of parent birds.

Genetic fingerprinting

Electrophoresis has many applications. DNA is highly specific so the bands produced using this process can help with identification. In some crimes DNA is left at the scene. Blood and semen both contain DNA specific to an individual. DNA evidence can be checked against samples from suspects. This is known as genetic fingerprinting. Genetic fingerprinting can be used in paternity disputes. Each band of the DNA of the child must correspond with *either* a band from the father or mother.

Isolating genes

Along chromosomes are large numbers of genes. Scientists may need to identify and isolate a useful gene; one way of doing this is to use the enzyme reverse transcriptase. This is produced by viruses known as retroviruses. Reverse transcriptase has the ability to help make DNA from mRNA.

In examinations you may experience questions on a wide range of applications.

(a) Someone with leukaemia must have a bone-marrow transplant. The DNA of the marrow cells after transplant must be the same as the donor if the operation has been successful.

(b) When breeding animals, the DNA of potential breeding pairs can be checked to make sure that they come from different families. Inbreeding is avoided.

Stage 1
When a polypeptide is about to be made at a ribosome, reverse transcriptase allows a strand of its coding DNA to be made.

mRNA UAA GCC GAU
single DNA ATT CGG CTA

Stage 2
The single stranded DNA is parted from the mRNA. single DNA ATT CGG CTA

Stage 3
The other strand of DNA is assembled using DNA polymerase.

DNA TAA GCC GAT
ATT CGG CTA

Using this principle the exact piece of DNA which codes for the production of a vital protein can be made.

Progress check

1 A length of DNA was prepared then electrophoresis was used to separate the sections. The statements below describe the process of electrophoresis but they are in the wrong order. Write the letters in the correct sequence.

A electrodes apply an electrical field
B DNA sections are put into a well in a slab of agar gel
C smaller pieces of DNA move more quickly down the agar track with larger ones further behind
D the gel and DNA are then covered with buffer solution which conducts electricity
E restriction endonucleases can be used to cut up the DNA

2 Reverse transcriptase is an enzyme which enables the production of DNA from RNA. Work out the sequence of organic bases along the DNA of the following RNA sequence.   AAUGCCCGGAUU

DNA₁ TTACGGGCCTAA
DNA₂ AATGCCCGGATT
2 RNA AAUGCCCGGAUU

1 E B D A C

6.4 Genetically modified organisms

After studying this section you should be able to:

● *outline the main features of a range of genetically modified organisms*
● *explain the advantages and disadvantages of using a range of genetically modified products*

LEARNING SUMMARY

Examples of genetic modification

AQA A	M2
AQA B	M2
OCR	M1
NICCEA	M1

Animals and plants can also have their DNA changed. A new gene can be added to give the organism a new property or feature.

Genetically modified soya bean plants

Gene transfer can be achieved in a number of ways; e.g. *Agrobacterium tumefaciens* specialises in invading plants through roots. It sends a plasmid into the host cell which becomes incorporated into the host cell's chromosomes. The genetic engineer puts a new gene into the plasmid which takes the gene into the cell. This results in a new feature.

In the USA large quantities of soya beans are produced. Selective herbicides (weed killers) are effective against broad-leaved plants. They could not be used in soya fields because the crop is also broad-leaved. Farmers needed to use expensive mechanical methods to kill the weeds. Genetic engineers transferred a gene using a **vector** into a soya bean plant which gave resistance to selective herbicides. The vector was a bacterium which entered the soya bean plant taking the useful gene with it. Since then the modified soya seeds have been grown all over the world. Farmers can use selective herbicides in their soya fields and keep them weed free.

Genetically modified potato plants

A gene which allows the production of insecticide has been transferred into a new breed of potato plant. Aphids which feed on the sap of the plants take in the insecticide which kills many of them. It is not 100% effective so there are many resistant aphids in the environment. Ladybirds are natural predators which eat the aphids. It has been found that ladybird fertility has decreased, and they live half as long as normal. In the long term the aphid pests may increase even more, becoming a greater problem. Ladybirds could even become extinct. Gene transfer may have problems in store for the human race.

Ethical issues

AQA A	M2
AQA B	M2
OCR	M1
WJEC	M1
NICCEA	M1

The applications of gene technology have huge implications. People must assess the advantages against the potential dangers.

In the examination you may be asked to consider a scenario involving genetic engineering. You will not be credited for simply stating that you agree or disagree. Show that you have a balanced view giving advantages *and* disadvantages. Show awareness of the consequences which may take place. Give details of possible sequential effects.

● It is possible to locate a defective gene in a fetus. The gene may lead to a condition such as muscular dystrophy. Consider potential action and the responsibility of knowing this information.

● Companies now are able to change a species drastically to produce something useful to humans. They even patent the new life form so that its reproduction is under their licence. Is it moral to change a species?

● If potatoes contain insecticide which kills aphids and their ladybird predators, then what effect may it have on human consumers of the genetically modified potatoes?

● If genetically modified soya plants resistant to herbicides breed with weeds, they may pass this feature on. What effect would this have on crop yields?

Life is based on DNA. Different species possess DNA with different sequences of organic bases. Adding new sections to give new properties has advantages but should be handled with care. There may be consequences.

Sample questions and model answers

The table below shows some mRNA codons and the amino acids which are coded by them.

	second position					
	U	C	A	G		
first position U	Phe	Ser	Tyr	Cys	U	third position
	Phe	Ser	Tyr	Cys	C	
	Leu	Ser	stop	stop	A	
	Leu	Ser	stop	Trp	G	

Key to amino acids

Ser – serine Tyr – tyrosine
Phe – phenylalanine Trp – tryptophan
Leu – leucine Cys – cysteine

Use the information in the table to help you answer the following questions.

1

(a) Give a sequence of mRNA bases which would code for leucine. [1]

UUA or UUG

(b) What does the mRNA base sequence UAC code for? [1]

Tyrosine

2

The mRNA sequence UCA codes for serine. Work out the base pairs on the DNA. [3]

UCA is coded for by these bases, AGT
AGT links to the bases TCA
So the DNA is AGT
                    TCA

3

Use evidence from the table to show that serine is an example of the degenerate code. [1]

It is coded for by four different base sequences.

4

UAG codes for 'stop'. Explain the effect of the 'stop' code during the process of protein synthesis. [2]

It is responsible for the polypeptide being terminated which allows it to leave the tRNA once all the amino acids have been linked.

5

During protein synthesis,

(i) where does translation take place?

At a ribosome

(ii) where does transcription take place? [2]

In a nucleus

Practice examination questions

Try all of the questions and check your answers with the mark scheme on page 141.

1 The two DNA sequences below are cut by the enzymes *Eco*R1 and *Hin*d111.

```
G │ A A T T C           A │ A G C T T
C T T A A │ G           T T C G A │ A
```
  *Eco*R1   *Hin*d111

Two identical pieces of DNA were cut with each enzyme, separately. The resulting pieces are shown below.

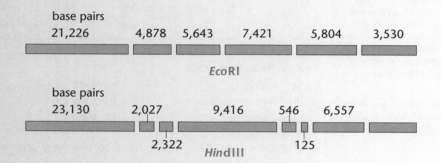

base pairs
21,226   4,878   5,643   7,421   5,804   3,530

*Eco*RI

base pairs
23,130   2,027   9,416   546   6,557

2,322   125
*Hin*dIII

(a) How many times did the base sequence cut by *Eco*R1 occur along the DNA? Give a reason for your answer. [2]

(b) What term is given to each end of a piece of cut DNA? [1]

(c) How many base pairs were there on the final piece of DNA cut by *Hin*d111? Show your working. [2]

(d) Describe how the use of the enzymes *Eco*R1 and *Hin*d111 are useful in genetic fingerprinting. [3]

2 The diagram below shows a stage in the process of mitosis.

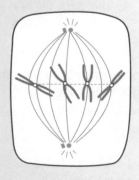

(a) Give the stage of mitosis shown. [1]

(b) How many chromosomes would there be in the daughter cells? [2]

3 (a) Explain how a useful gene can be transferred from a human pancreas to a bacterium. [6]

(b) Describe how the transgenic bacteria could be cultured in an industrial fermenter to produce a useful product. [5]

(c) Give ONE example of a hormone which is produced as a result of gene transfer from pancreas cells. [1]

Practice examination questions *(continued)*

4 The table below shows the relative organic base proportions found in human, sheep, salmon and wheat DNA.

Organism	Proportion of organic bases in DNA (%)			
	Adenine	*Guanine*	*Thymine*	*Cytosine*
human	30.9	19.9	29.4	19.8
sheep	29.3	21.4	28.3	21.0
salmon	29.7	20.8	29.1	20.4
wheat	27.3	22.7	27.1	22.8

(a) Refer to the proportion of organic bases in salmon DNA to explain the association between specific bases. [2]

(b) Suggest a reason for the small difference in proportion of the organic bases adenine and thymine in sheep. [1]

(c) All species possess adenine, guanine, thymine and cytosine in their DNA. Account for the fact that each species is different. [2]

5 It was suspected that a person had taken an egg from the nest of a rare bird. DNA samples were taken from the egg and both parent birds. The DNA profiles shown below were made using electrophoresis.

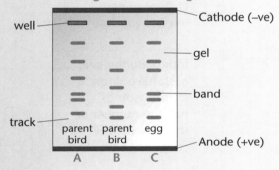

a view looking down on the agar slab

(a) Which type of enzyme is used to cut DNA before electrophoresis? [1]

(b) Do the electrophoresis results suggest that the egg was taken from the nest? Give a reason for your answer. [1]

(c) Suggest TWO other applications for electrophoresis. [2]

6 A new genetically modified soya bean plant has been developed. It has a new gene which prevents it from being killed by herbicide (weed killer).

(a) Describe the stages which enable a gene to be transferred from one organism to another. [5]

(b) Explain how the genetically modified soya plants result in higher bean yields. [3]

(c) What would be the danger if the genetically modified soya plants interbred with the plants of the hedgerows? [3]

(d) Suggest why people may object to the growing of genetically modified soya bean plants. [2]

Chapter 7

Continuity of life

The following topics are covered in this chapter:

- *Variation*
- *Plant reproduction*
- *Human reproduction*

7.1 Variation

After studying this section you should be able to:

- *understand the significance of meiosis to sexual reproduction*

LEARNING SUMMARY

Why is meiosis so important in sexual reproduction?

AQA A	M2
AQA B	M2
OCR	M1
WJEC	M1
NICCEA	M1

When sexual reproduction takes place the male and female must produce gametes (sex cells). These are produced by the process of meiosis. This is sometimes called reduction division because a **diploid** parental cell divides to produce four **haploid** cells.

- parental cell has chromosomes in pairs (**diploid**)

> Note that the diploid number in a human body cell is 46 (23 pairs of chromosomes).

- four daughter cells each have single **chromosomes**, half the number of the parental cell (**haploid**)

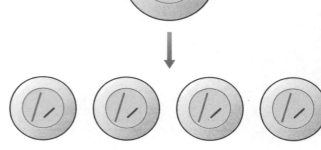

All cells produced by meiosis are **different** to the parental cell and to each another. This is highly significant and is a major factor in the **genetic variation** within a species. Consider human reproduction as an example. Every person inherits a set of chromosomes from the mother (23) and a set from the father (23). The matching chromosomes of each corresponding pair have the same gene in the same position (locus) along their length. They may be exactly the same gene but can be different **alleles**.

> Note that the haploid number in a human body cell is 23 (single) chromosomes.

An allele is a different expression of a gene. Each pair of alleles along homologous chromosome pairs can be homozygous, e.g. **AA**, **aa** or heterozygous, e.g. **Aa** (where **A** is **dominant**, and **a** is **recessive**).

How are genetically different cells produced during meiosis?

> The key event in producing genetically different gametes takes place during the first stage of meiosis – **prophase 1**. The diagrams on the next page show the process of **crossing over** of a pair of homologous chromosomes and the consequences this has on producing **different allele combinations**.

KEY POINT

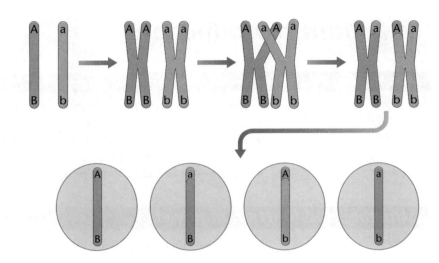

If this diagram represents one cross-over or chiasma then imagine what happens when there are **many** cross-overs along a pair of chromosomes. Each person has 23 pairs of homologous chromosomes. What a lot of cross-overs!

Note that the pair of homologous chromosomes shown have different alleles.

A represents an allele dominant to a, a recessive allele.

B represents an allele dominant to b, a recessive allele.

The chromosomes would have many more genes than merely ones shown by A, a, B, b. The diagram above shows the consequence of only one cross-over or chiasma. Cross-overs cause a difference in the combinations of alleles along a chromosome. The cross-over, as shown above, results in chromosomes with AB, Ab, aB and ab combinations.

With **many cross-overs** taking place along **all 23 pairs of chromosomes** it is not surprising that every cell produced by meiosis is genetically different.

Meiosis takes place by two divisions. Consider these two divisions in relation to the cross-over above.

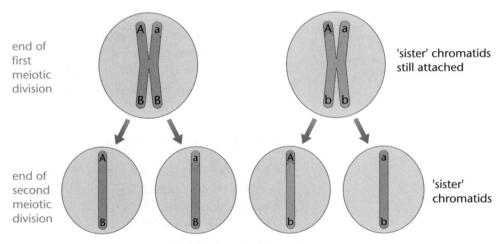

end of first meiotic division

'sister' chromatids still attached

end of second meiotic division

'sister' chromatids

The combination of each of these chromosomes with the others, results in further genetic variation.

Humans are given as examples of the consequences of a meiotic division. Plants and other organisms divide by **exactly the same principles**. Only the chromosome numbers and alleles differ.

The combination of each of these chromosomes with 22 others, results in further genetic variation; the random segregation of chromosomes. Human gametes are produced by meiosis. Only 1 homologous pair is shown in the diagram but there would be 22 other pairs. If each gamete is different, then the male gamete fusing with a female gamete is yet another source of variation.

7.2 Plant reproduction

After studying this section you should be able to:

- describe the structures and functions of flowers
- understand the development of pollen and ovules
- understand the significance of cross-pollination
- describe typical seed structure
- describe the special features of wind- and insect-pollinated plants

Structure and function of flowers

EDEXCEL ▶ M2

Sexual reproduction in plants takes place as a result of the activity of specific parts of flowers. The diagram below shows a section through a flower.

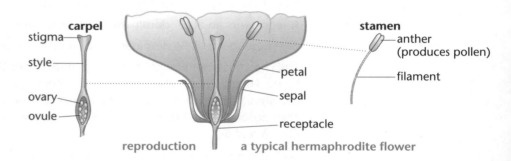

reproduction   a typical hermaphrodite flower

The functions of the parts are as follows.

- Sepals – collectively known as the calyx, protect the flower whilst in bud.
- Petals – if they are large and brightly coloured attract insects for the process of pollination.
- Stamens – consist of anther and filament. The anthers produce the male gametes (pollen) each of which contains a haploid set of chromosomes. The anther does this by meiotic division.
- Carpels are made of stigma, style and ovary. In the ovary are the female gametes (ovules) each of which also contains a haploid set of chromosomes.

> Candidates often confuse the male and female structures. Learn the structure names and functions carefully.

How does the anther make pollen?

The complete development takes place in a pollen sac inside an anther.

- A diploid pollen mother cell divides by meiosis.
- Each division forms 4 pollen grains.
- The haploid nucleus of each pollen grain divides by mitosis to form 2 haploid nuclei per pollen grain.
- One nucleus is the tube nucleus and one is the generative nucleus.
- The male gamete is contained inside the pollen grain.
- The outer cellular layer of the pollen grain is smooth for wind-pollinated flowers and rough for insect-pollinated flowers.

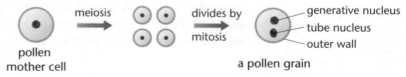

formation of a pollen grain

How does the ovary make ovules?

The complete development of each ovule takes place beginning with an embryosac mother cell.

- The embryosac mother cell (diploid cell) divides by meiosis to form 4 haploid cells.
- 3 of them degenerate.
- The nucleus of the 1 haploid cell remaining divides by mitosis.
- This produces 2 haploid nuclei, which again divide by mitosis.
- This produces 4 haploid nuclei, which again divide by mitosis.
- Finally this produces 8 haploid nuclei, and the structure is known as a mature embryosac.

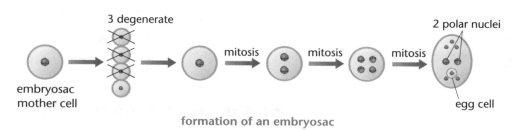

formation of an embryosac

Pollination

This is the movement of pollen from an anther to a stigma. The stigma often has a surface covered with a sticky sugary substance, acting as a super-glue to hold onto pollen grains.

Fertilisation

What is fertilisation?
This takes place during sexual reproduction. The male gamete fuses with a female gamete to produce a zygote. This first diploid zygote then divides many times by mitosis to produce a new organism.

Pollen reaches a stigma and sticks to the surface, the following events then take place.

- A pollen tube begins to grow from the pollen grain.
- The proteins for this tube are produced with the help of the tube nucleus which remains near the tip of the tube.
- The generative nucleus divides into 2 male nuclei which follow down the tube.
- The pollen tube grows into an ovule.
- One male nucleus fuses with the egg cell forming a diploid zygote; this divides many times to form the embryo of the seed.
- The other male nucleus fuses with the two polar nuclei forming a triploid cell; this divides many times to form the endosperm (food store) of the mature seed.
- This is double fertilisation because of the fusion to form the diploid zygote and the triploid primary endosperm cell.

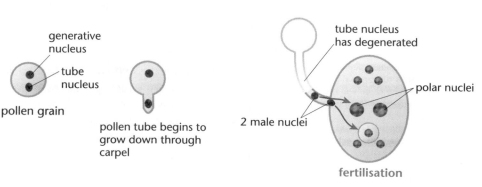

fertilisation

Seed structure

EDEXCEL M2

After fertilisation the flower parts die off leaving the ovary surrounded by its wall. A variety of different types of fruit form after this depending on the species of plant, e.g. in a plum the ovary wall becomes fleshy and swells. This protective outer wall is now known as the pericarp. A typical seed has an outer protective covering known as the testa.

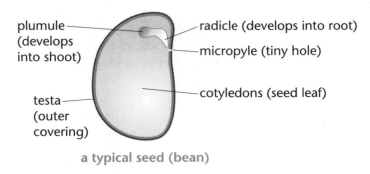

plumule (develops into shoot)

radicle (develops into root)

micropyle (tiny hole)

testa (outer covering)

cotyledons (seed leaf)

a typical seed (bean)

What special adaptations do plants have for reproduction?

There are many different structures and mechanisms which help plants to reproduce successfully. Each has evolved over a considerable period.

Many flowers are able to self-pollinate but there is an increased probability of disadvantageous alleles being exhibited in the phenotype of offspring.

> Cross-pollination is better as it increases the probability of advantageous alleles, strengthening the gene pool of a species.
>
> **KEY POINT**

Mechanisms of pollination

Plants have evolved a number of adaptations to make sure that self-pollination does **not** take place before pollen is transferred from one flower to another.

Protandry

This means that the anthers ripen first, to produce pollen before the carpel has ovules ready for fertilisation. Reproduction can only take place by pollen being taken to another flower, e.g. Rosebay Willowherb (*Chamaenerion angustifolium*). Correspondingly, the later-ripening ovules can only receive pollen from another flower.

Protogyny

Note that protandry and protogyny are different mechanisms, but the outcome is similar, i.e. they ensure that cross-pollination takes place.

This means that the stigmas ripen first, to produce ovules before the anther has pollen ready for fertilisation. Reproduction can only take place by pollen being received from another flower, e.g. Bluebell (*Endymion non-scriptus*). Correspondingly, the later-ripening anthers can only produce pollen capable of pollinating another flower.

Dioecious plants

Every female holly tree needs a male pollinator, otherwise there will be no berries! Holly flowers are dull and insignificant. Do you think they are insect- or wind-pollinated?

These are plant species which have a separate plant which produces just male flowers and another which produces exclusively female ones, e.g. Holly (*Ilex* species). This means that cross-pollination must take place if reproduction is to be successful. In a garden two holly trees, a male and a female, must be planted if reproduction is to be successful. Just the female alone will not successfully produce berries. It is possible that a single holly plant produces berries but there is always a male plant close by!

Features of wind-pollinated and insect-pollinated plants

The transport of pollen from one plant to another requires some mechanism to carry the pollen since, unlike a sperm, it is non-motile. The table below shows the typical features of wind- and insect-pollinated plants.

Features	
Wind-pollinated plants	*Insect-pollinated plants*
Petals are small or absent so that anthers and stigmas are exposed to the wind	Petals are large to attract insects
Stamens hang out of the flower so that pollen can be blown away on the air currents. Long styles allow the stigma to be exposed to catch the pollen blown by the wind	Stamens and stigmas are found inside the flower. Petals often in a tubular arrangement. This forces the insect to brush against anthers for pollen pick up and brush against stigmas which releases pollen.
If petals are present they are are often green	Petals are coloured to attract insects
No scent produced	Scent often produced, e.g. perfumed roses or the stinking-flesh scent of *Fritillaria*
No nectaries are present	Nectaries are present which contain (nectar) carbohydrate as an attractant for insects. They are deep in the receptacle of the flower, increasing the probability of the insect brushing against anther and stigma.
Flowers are well above the leaves, exposed to air currents, often found on tall stems	Many flowers are at a lower level, e.g. pansy
Very high pollen quantity produced which increases the probability of successful pollination	Lower quantity of pollen produced because insect pollination has a high chance of success
Pollen smooth and aerodynamic	Pollen rough, enabling it to cling to insect bodies

Wind-pollinated plants have flowers which are hardly noticeable. The diagram below shows a typical grass flower.

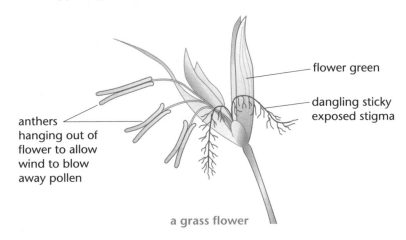

anthers hanging out of flower to allow wind to blow away pollen

flower green

dangling sticky exposed stigma

a grass flower

Progress check

Explain the function of protandry and protogyny in flowering plants.

Protandry ensures that the pollen ripens first and protogyny causes the ovules to ripen before the pollen. Both ensure that self-pollination does not take place.

7.3 Human reproduction

After studying this section you should be able to:

- *describe the structure and functions of male and female reproductive systems in humans*
- *describe the roles of the hormones which control the menstrual cycle*
- *describe the transfer of gametes leading to fertilisation in humans*
- *outline implantation and fetal development in humans*
- *understand the process of birth and lactation*

The reproductive structures

AQA A M2
EDEXCEL M2

When a baby is born it already has the structures which can eventually result in the production of offspring. Once adolescence is reached the sex organs of a person begin to function, as they mature.

The male reproductive system

The diagram below shows the male reproductive system.

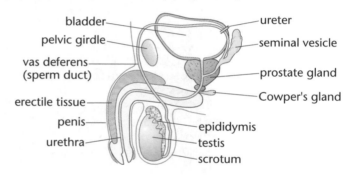

bladder
pelvic girdle
vas deferens (sperm duct)
erectile tissue
penis
urethra
ureter
seminal vesicle
prostate gland
Cowper's gland
epididymis
testis
scrotum

What are the functions of the male reproductive structures?

- The key structures are the two **testes** in which the sperm develop. Precisely, it is in the many **seminiferous tubules** where the millions of sperm are formed.
- The seminiferous tubules all lead to the **epididymis**, a coiled tube. Mature sperm are stored in the epididymis which in turn leads to the **vas deferens** (sperm duct).
- Sperm move through the vas deferens during ejaculation – a muscular spasm by which sperm are ejected from the male reproductive system.
- Each sperm is equipped with a tail which enables it to swim to the female gamete. The sperms need fluid to swim in, which they acquire on their journey from three structures:

  (a) prostate gland
  (b) Cowper's gland
  (c) seminal vesicle.

  These secrete alkaline fluid which aids sperm survival, but additionally the seminal vesicle produces sucrose, an energy source for the sperm.
- The sperm and fluid together are known as semen.
- Semen leaves the body via the penis. Here it passes through the urethra, a central tube in the penis which is shared with urine excretion, although not at the same time!
- The penis can become firm and erect by the inflow of blood in the erectile tissue around the urethra.

You will probably remember a lot of this information from GCSE. Build on what you already know. Additional facts about the glands, vesicles and vas deferens will widen your knowledge.

- A sac, the scrotum, covers each testis. This gives limited protection and helps keep the testes 2°C below the normal body temperature. Sperm need this lower temperature to develop. They must not overheat!
- The testes also produce the hormone **testosterone**, which helps develop secondary sexual characteristics.

The female reproductive system

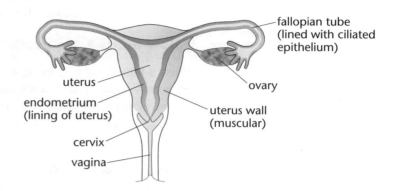

What are the functions of the reproductive structures?

- The key structures are the two **ovaries** which produce hormones and develop the ova (female gametes).
- A **fallopian tube** connects each ovary to the uterus. Each fallopian tube is a muscular tube lined with **ciliated epithelia** and **glandular** cells which secrete mucus.
- The action of the cilia of the cells lining the fallopian tube moves the ovum progressively towards the uterus after **ovulation**.
- The mucus is propelled along by the **cilia**, so that the ovum moves smoothly **without dehydration**.
- The uterus is a cavity lined by a thick layer of smooth muscle. This will, by contractions, help the mother to give birth in the final stage of pregnancy.
- The uterus is lined by the **endometrium**, a mucous membrane which has a rich supply of capillaries. This is used to supply the embryo with services during pregnancy.
- The vagina is a muscular tube which is connected to the uterus via the narrow 'neck' of the uterus, the **cervix**.

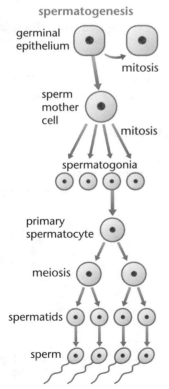

How are the sperms produced in the testes?

The process is called **spermatogenesis**. The sequence of this process takes place as follows:

- cells from the germinal epithelium on the outside of the seminiferous tubules divide by mitosis
- this produces sperm mother cells (spermatogonia)
- each sperm mother cell divides by mitosis to produce many more identical cells
- each spermatogonium grows to form a primary spermatocyte which divides by the first meiotic division to form secondary spermatocytes
- each secondary spermatocyte then completes the second meiotic division to form spermatids
- Sertoli cells are a source of nutrition for the spermatids; the heads grow a tail and become mature sperms after feeding at these cells
- finally they fall away to be transported to the epididymis for storage.

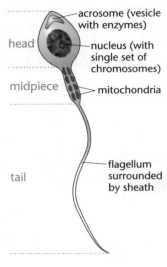

acrosome (vesicle with enzymes)

head

nucleus (with single set of chromosomes)

midpiece

mitochondria

tail

flagellum surrounded by sheath

The structure of a human sperm

- The head consists of a haploid nucleus, carrying a single set of chromosomes from the male. Near the front of the head is the acrosome, a package of proteolytic enzymes which enable the sperm to pass through the protective layer around an ovum.
- The mid-piece is full of mitochondria which release energy necessary to fuel the life processes of the sperm on its journey.
- The tail consists of a sheath around a flagellum. Much of the energy released in the mid-piece powers the swimming action of the flagellum.

How are the ova produced in an ovary?

When a female baby is born, each ovary contains thousands of primary follicles. All remain dormant until adolescence begins. At maturity it is usual that one follicle will develop each month and produce an ovum.

The process is by oogenesis. The sequence of this process begins as a female fetus is developing in a uterus.

Stage 1 (foetus to birth)

- Fetal cells from the germinal epithelium on the outside of its ovaries divide by mitosis to form oogonia.
- Each oogonium grows to form a primary oocyte.
- Each primary oocyte becomes surrounded by a layer of follicle cells. It is then known as the primary follicle (Graafian follicle).
- A female baby has thousands of primary follicles at birth, but all remain dormant until adolescence begins.

Stage 2 (maturity to menopause)

- At maturity, during each month it is normal for just one primary follicle to develop to maturity stimulated by a hormone 'trigger' (see page 104).
- The primary oocyte divides by the first meiotic division to form a secondary oocyte and a smaller polar body.
- A second meiotic division takes place to form the large secondary oocyte and much smaller second polar body.
- The secondary oocyte is known as the ovum as it is released from the primary follicle.
- The release of the ovum is known as ovulation.
- The ruptured primary follicle still has an important role! It changes into a corpus luteum (yellow body) which produces an important hormone – progesterone.

The ovary shown below shows the events leading to ovulation.

oogenesis

germinal epithelium

mitosis

mitosis

oogonia

primary follicle

primary oocyte

first polar body

secondary oocyte

second polar body

secondary oocyte

The diagram of the ovary shows the events which take place from the development of the primary oocyte to the degeneration of a corpus luteum. The diagram shows the sequence for one primary follicle in its development. All stages are not present at the same time! The sequence shown would only be visible if a time-lapse sequence followed the **same** follicle during development.

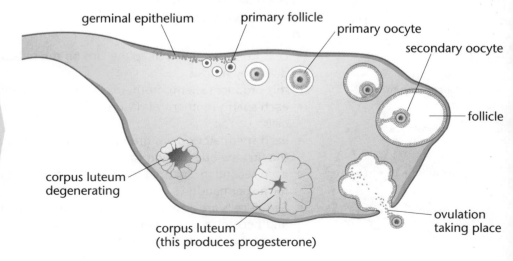

germinal epithelium   primary follicle

primary oocyte

secondary oocyte

follicle

corpus luteum degenerating

corpus luteum (this produces progesterone)

ovulation taking place

The events leading to fertilisation

During copulation the erect penis is inserted into the vagina. When ejaculation takes place the semen is released at the cervix. The semen immediately coagulates so that the female retains the sperm. Within a few minutes the sperm then begin to swim in a fluid produced by the female. Fertilisation normally takes place in the fallopian tube, where one sperm fuses with an ovum. The diagram below shows fertilisation.

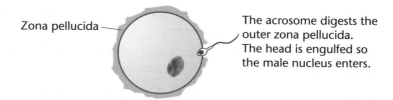

Zona pellucida

The acrosome digests the outer zona pellucida. The head is engulfed so the male nucleus enters.

> Remember that the ovum can also be called the secondary oocyte.

- Many sperm are attracted to the ovum by chemotaxis.
- The acrosome, a vesicle containing hydrolytic enzymes, breaks down part of the zona pellucida to allow the sperm entry into the secondary oocyte.
- The cell membrane of the sperm head fuses with the cell membrane of the secondary oocyte.
- The sperm nucleus is engulfed and moves into the cytoplasm of the secondary oocyte.
- Finally the two sets of haploid nuclei fuse together to form the diploid nucleus of the zygote, which may go on to produce the fetus.

What controls the menstrual cycle?

Events which take place within a female, such as ovulation and menstruation, need to be coordinated. Menstruation, the break down and loss of the lining of the uterus, would have a devastating consequence if a woman was pregnant. The foetus would be miscarried. The menstrual cycle is controlled by the secretion of hormones by the endocrine system. The flow diagram below shows how the process achieves coordination.

> Note that at the start of the menstrual cycle FSH is supported by a smaller quantity of LH to stimulate the development of a follicle. Later the proportion reverses so that a larger amount of LH is aided by a smaller amount of FSH resulting in ovulation.

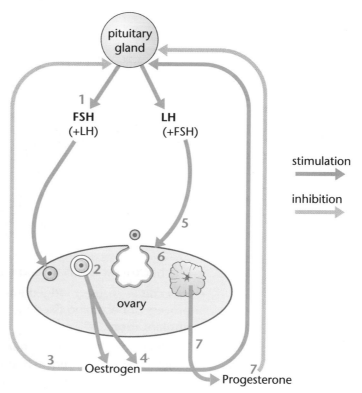

Stages of the menstrual cycle

1 (a) FSH (follicle stimulating hormone) is secreted into the bloodstream by the pituitary gland. It stimulates the development of a Graafian follicle and so triggers the development of an ovum.

   (b) At the same time a small amount of LH (luteinising hormone) is secreted by the pituitary gland which reinforces the effect of FSH.

2 As a follicle develops, its wall (**theca**) begins to secrete oestrogen which stimulates the building of the endometrium (lining of uterus).

3 The oestrogen inhibits the secretion of FSH *temporarily* but LH secretion continues.

4 A peak of oestrogen is reached which results in a surge of LH with some FSH which is no longer inhibited.

5 When LH peaks it causes ovulation – the follicle ruptures releasing the ovum.

6 The empty follicle now changes role and becomes a corpus luteum which begins to secrete progesterone.

7 (a) Progesterone keeps the endometrium in position, as it will be needed if a foetus is to develop in the uterus.

   (b) Progesterone also inhibits the secretion of any FSH or LH by the pituitary gland. Ovulation is ultimately prevented by high concentrations of progesterone.

8 If no sperm fertilises an ovum during the cycle then the corpus luteum degenerates and a drop in progesterone takes place. Low progesterone does not inhibit the FSH and LH so that they are both able to be secreted again.

The cycle is now complete – *GO BACK TO STAGE 1!*

> Remember that all hormones are secreted into the bloodstream.

Pregnancy

During fertilisation the **zygote** is produced. This **diploid** cell is moved down the **fallopian tube** by the **cilia**. On this journey it undergoes cell division by mitosis. It takes about four days to reach the uterus by which time it has become a **ball of cells (blastocyst)**.

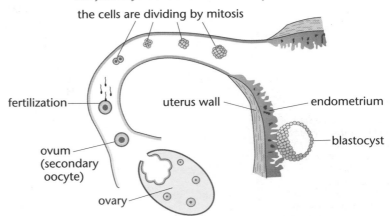

the journey: from ovulation to implantation

the cells are dividing by mitosis

fertilization — ovum (secondary oocyte) — ovary — uterus wall — endometrium — blastocyst

- The **blastocyst** is hollow and ready to become attached to the uterus.

- It becomes securely lodged into the **endometrium (lining of uterus)**. This process of becoming fixed into the surface of the endometrium is known as **implantation**.

- The blastocyst secretes the hormone **human chorionic gonadotrophin (HCG)** which **prevents degeneration** of the **corpus luteum** so that **progesterone** secretion continues into pregnancy.

- Cells of the blastocyst divide further resulting in the development of a placenta.
- The placenta secretes oestrogen and progesterone. Together they maintain the endometrium and prepare the mammary glands for lactation (milk production).
- The placenta also secretes relaxin. This relaxes the elastin fibres joining the bones of the pelvic girdle together and helps to expand the cervix during the final stages of pregnancy.

Further development of the placenta

After the blastocyst implants into the endometrium it has access to a range of nutrients and oxygen as well as the ability to allow waste products such as carbon dioxide to diffuse away into the mother's blood. All this is possible because the outer layer of cells of the blastocyst form the chorionic villi, finger-like projections of fetal origin which embed into the endometrium. They are surrounded by spaces filled by the mother's blood (maternal blood spaces). The combination of the chorionic villi and maternal blood spaces gives a high surface area enabling the exchange of chemicals between embryo and mother.

- Chemicals which diffuse from mother to fetus include, oxygen, glucose, water, amino acids, fatty acids, glycerol, minerals, vitamins, some hormones and some antibodies.
- Chemicals which diffuse from embryo to mother include urea and carbon dioxide as well as some hormones and water.
- Harmful chemicals including drugs and alcohol and even some viruses can cross the placenta into the foetus. Fetal development can be impaired.
- The umbilical artery takes waste substances to the placenta for excretion and the umbilical vein collects useful substances from the placenta and transports them to the fetus.

Remember that the chorionic villi have the fetal nuclei, whereas the endometrium is parental, having identical nuclei to the mother.

Birth

The 40-week period during which time the fetus develops in the uterus is known as gestation. The final stage involves birth (parturition) as follows:

- the foetus moves into the 'head-down' position
- the cervix begins to dilate and eventually becomes wide enough to allow even the widest part of the baby (the head) to pass through

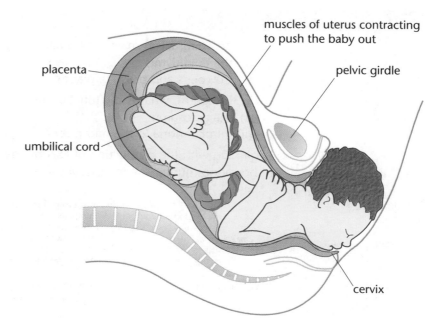

Many births in hospital are induced using oxytocin artificially.

- contractions of the uterus wall, (known as labour) are stimulated by the hormone oxytocin, secreted by the pituitary gland.
- another hormone, prostaglandin, is secreted by the placenta which increases the frequency and force of contractions of the uterus
- birth finally takes place as the baby is expelled from the uterus, closely followed by the umbilical cord and placenta.

Lactation

This is the production of milk and includes its ejection from the mammary glands to feed the baby. Preparation of these glands takes place during gestation due to the effect of oestrogen and progesterone. However, milk production cannot take place when the level of progesterone in the blood is high. In the final stages of gestation progesterone falls which initiates lactation. Prolactin, yet another hormone, is secreted by the pituitary gland which finally results in the mother lactating. The physical effect of the baby's sucking reflex at the nipples stimulates the lactation process to continue.

The first milk of the lactating mother is most important. It contains a high concentration of antibodies which gives the baby a degree of immunity from a range of diseases.

Progress check

1 (a) In the mammalian male reproductive system what is the function of:
    (i) the epididymis
    (ii) Cowper's and prostate glands?

   (b) In the sperm, what is the function of:
    (i) the acrosome
    (ii) mitochondria in the mid-piece?

2 Beginning with the production of FSH by the pituitary gland, describe the role of hormones during the menstrual cycle.

Corpus luteum degenerates, progesterone level falls and FSH can be produced again.
Progesterone inhibits the secretion of FSH, ovulation is prevented by high concentrations of this hormone.
progesterone. Progesterone keeps the endometrium in position, and is needed to develop the foetus.
pituitary gland; ovulation takes place; the follicle becomes a corpus luteum which begins to secrete
stimulates the building of the endometrium; oestrogen inhibits the secretion of FSH; a surge of LH secreted by
2 Primary follicle develops and begins the development of an ovum: oestrogen is secreted into the blood; this
to move.
(b) (i) contains enzymes to break down the ovum membrane for sperm entry; (ii) energy release for flagellum
1 (a) (i) storage of sperm; (ii) addition of fluid for sperm to swim and nutrients.

Sample question

(a) The diagram below shows a mammalian sperm cell.

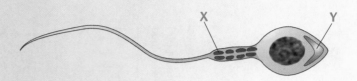

The diagram shows a sperm. Immediately you may think, 'This is like GCSE.'. The question builds on GCSE knowledge. Learn the structures which are 'new' to you.

   (i) Name each organelle labelled in the diagram and describe its function.

Organelle X   [2]

This is a mitochondrion; it releases energy by aerobic respiration, to move the flagellum.

Remember to learn all structures and their functions. Here you need to recognise the mitochondrian and relate the energy release to produce a swimming action.

Organelle Y   [2]

This is an acrosome, it is an enzyme package which helps the sperm penetrate through the zona pellucida and plasma membrane.

  (ii) What is the function of the nucleus?   [1]

Carries a haploid set of chromosomes.

(b) (i) Where in the testes are the sperm stored?   [1]

Epididymis.

  (ii) Give **two** functions of the seminal vesicles and Cowper's glands during ejaculation.   [2]

Fluid is added for sperm to swim in.
Nutrients are added to supply energy which enables sperm to swim.

(c) The graph below shows the levels of oestrogen and progesterone in the blood of a woman during one month.

Many concepts you learn may be examined using graphs. Always look for peaks and troughs which mark significant events.

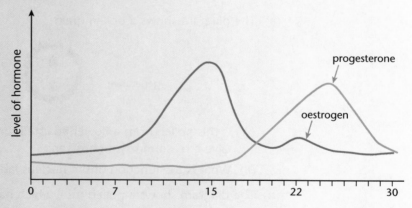

   (i) Which hormone, produced by the pituitary gland, would have resulted in the peak of oestrogen?   [1]

You would only score this mark about the peak of oestrogen if you could *work backwards* and link the hormone (FSH) to the stimulation of oestrogen.

FSH or follicle stimulating hormone.

  (ii) What evidence shown on the graph shows that the woman is not pregnant?   [1]

Progesterone level has fallen.

Practice examination questions

1 The graph shows the relative level of progesterone in the blood of a woman during one menstrual cycle.

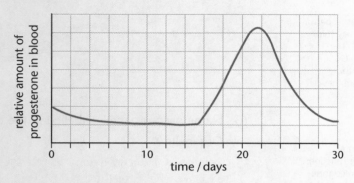

(a) (i) Precisely what produced the progesterone? [1]

    (ii) How does progesterone reach the uterus? [1]

(b) Was the woman pregnant? Give a reason for your answer. [1]

(c) The role of the progesterone is to maintain the endometrium (lining of the uterus). Suggest what would happen if no progesterone was produced during pregnancy. [1]

2 (a) Some plant species have adaptations for cross-pollination. Explain how each of the following species ensures that cross-pollination takes place.

    (i) Bluebell (*Endymion non-scriptus*) using protogyny

    (ii) White dead nettle (*Lamium album*) using protandry. [2]

(b) The flowers of the Conference pear tree (*Pyrus communis*) self-pollinate.

    (i) Explain **one** advantage to a gardener, of self-pollination. [1]

    (ii) Explain **two** disadvantages to the species of self-pollination. [2]

3 (a) The diagram shows a pollen grain.

    (i) Is this pollen from a flower adapted for wind or insect pollination? Give a reason for your answer. [1]

    (ii) What is the function of the nucleus labelled X? [1]

(b) The diagram shows a pollen tube which has just penetrated an embryosac.

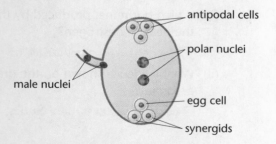

Describe the role of each male nucleus in fertilisation. [4]

Chapter 8

Energy and ecosystems

- Energy flow through ecosystems
- Energy transfer and agriculture
- Nutrient cycles

- Colonisation and succession
- Effects of human activity on the environment

8.1 Energy flow through ecosystems

After studying this section you should be able to:

- outline the process of photosynthesis
- identify the biotic and abiotic factors of an ecosystem
- understand the roles of producers, consumers, and decomposers
- understand the flow of energy through an ecosystem

LEARNING SUMMARY

Photosynthesis

EDEXCEL   M3
WJEC   M2
NICCEA   M2

Before energy is available to organisms in an ecosystem photosynthesis must take place. This is the process by which green plants make carbohydrates. The main stages take place in the chloroplasts.

Photosynthesis is often summarised as the production of glucose and oxygen, from carbon dioxide and water, with the help of light and chlorophyll. The true details of the process are much more complex than this!

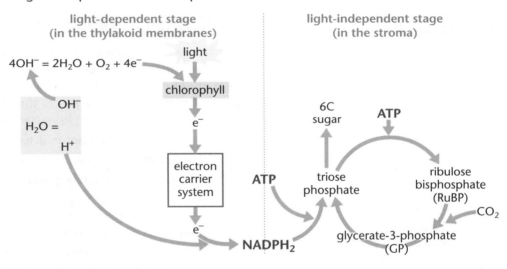

Light-dependent reaction

- Light energy results in the excitation of electrons in the **chlorophyll**.
- These electrons are passed along a series of electron acceptors collectively known as the **electron carrier system**.
- Energy from excited electrons funds the production of **ATP** (adenosine triphosphate).
- The final electron acceptor is **NADP⁺**.
- Electron loss from chlorophyll causes the splitting of water (photolysis).

Look out for the useful substances produced by the light-dependent stage. **ATP** and **NADPH₂** are regularly needed in answers!

$$H_2O = H^+ + OH^- \quad \text{then} \quad 4OH^- = 2H_2O + O_2 + 4e^-$$

- Oxygen is produced, water to re-use, and electrons stream back to replace those lost in the chlorophyll.
- Hydrogen ions (H^+) from photolysis, together with $NADP^+$ form **NADPH₂**.

Light-independent reaction

- Two useful substances are produced by the light-dependent stage – ATP and $NADPH_2$.
- They react with glycerate-3-phosphate (GP) to produce a triose sugar – triose phosphate.
- Triose phosphate is used to produce a 6C sugar.
- Some $NADPH_2$ is used, together with ATP to form ribulose bisphosphate (RuBP).
- RuBP together with carbon dioxide form more GP to complete the Calvin cycle.

Progress check

(a) Where in a chloroplast do the following take place:
  (i) the light-dependent stage
  (ii) the light-independent stage?

(b) At the end of the light-dependent stage, which **two** substances are produced that are needed for the light-independent stage?

(c) When does the light-independent stage take place?

(d) What is the function of carbon dioxide in the light-independent stage?

(d) Used in the reaction to convert ribulose bisphosphate into glycerate-3-phosphate.
(c) Immediately after the light-dependent stage (Never state in the dark!).
(b) ATP and $NADPH_2$, or reduced NADP.
(a) (i) thylakoid membranes (ii) stroma.

What is an ecosystem?

EDEXCEL — M3
OCR — M1
WJEC — M2
NICCEA — M2

The study of ecology investigates the inter-relationships between organisms in an area and their environment. The importance of photosynthesis to all organisms of an ecosystem must be considered. The plants (producers) make carbohydrates and are the source of most energy available to the organisms of an ecosystem. Before explaining the term ecosystem some important terms need to be defined:

- habitat is the area where an organism lives
- population is the number of organisms of one species living in an area
- community is a number of different populations living in an area
- biotic factors are factors caused by living organisms which influence other organisms in their environment, e.g. plants being consumed by herbivores or one species predating upon another
- abiotic factors are non-living factors which influence organisms in their environment, e.g. pH of the soil or the temperature of the environment
- niche is the precise way in which an organism fits into its environment and what it does there, e.g. a fish may survive within a temperature range of 30°C –35°C, and a pH range of 5–8 and eat a specific type of plant. All of its specific requirements for life are its niche
- competition is where different organisms occupy a similar niche, e.g. slugs and snails living in a garden both consume lettuce leaves.

Green plants are also known as photo-autotrophs. An autotroph is an organism which takes in simple inorganic chemicals and assembles more complex organic chemicals. Some of these can be later respired to release energy needed for life.

An ecosystem is a distinctive and stable ecological unit in an area and consists of the following features:

- different populations of organisms living and interacting together within a community
- all abiotic factors of their environment
- the energy flow through food chains and webs
- the cycling of nutrients to be re-used by the community.

Note that the term ecosystem is difficult to define! Make sure that you remember all four parts of this definition.

There may be no physical barrier between one ecosystem and the next, e.g. a desert ecosystem may exist alongside a tropical ecosystem, which has much more rainfall. Each specific ecosystem is self-sustaining and relies on the cycling of nutrients and special adaptations of the component organisms.

Some organisms may move from one ecosystem to another, e.g. a dragonfly larva is part of a food web in a pond, but after reaching the maturity of adulthood it flies into a terrestrial ecosystem. Also a migratory bird such as the insectivorous swallow flies from Britain to South Africa to avoid the winter but makes the return journey for summer. In this instance it occupies a similar niche in both countries.

Energy flow though an ecosystem

Sunlight energy enters the ecosystem and *some* is available for photosynthesis. *Not all* light energy reaches photosynthetic tissues. Some totally misses plants and may be absorbed or reflected by such items as water, rock or soil. Some light energy which does reach plants may be reflected by the waxy cuticle or even miss chloroplasts completely!

> Around 4% of light entering an ecosystem is actually used in photosynthesis.

 KEY POINT

The green plant uses the **carbohydrate** as a first stage substance and goes on to make **proteins** and **lipids**. Plants are a rich source of nutrients, available to the herbivores which eat the plants. Some energy is not available to the herbivores for two reasons:

1 green plants **respire** (releasing energy)
2 **not all parts** of plants may be **consumed**, e.g. roots.

Food chains and webs

Energy is passed along a food chain. Each food chain always begins with an autotroph (producer) then energy is passed to a primary consumer, then secondary consumer, then tertiary consumer and so on.

direction of energy flow →

Producer → primary consumer → secondary consumer → tertiary consumer
(herbivore)   (1st carnivore)   (2nd carnivore)

The following example shows four food chains linked to form a food web.

Note that a small bird is a secondary consumer when it eats apple codling moths but a tertiary consumer when it eats greenfly.

The producers always have more energy than the primary consumers, the primary consumers more than the secondary consumers and so on, up the food web. Energy is released by each organism as it respires. Some energy fails to reach the next organism because not all parts may be eaten.

hawk
small bird
thrush   ladybird
snail   greenfly   apple coddling
                                    moth
lettuce   apple

Each feeding level along a food chain can also be represented by a **trophic level**. The food chain on the next page is taken from the food web above and illustrates trophic levels.

Energy may be used by an organism in a number of different ways:

• respiration releases energy for movement or maintenance of body temperature, etc.
• production of new cells in growth and repair
• production of eggs.

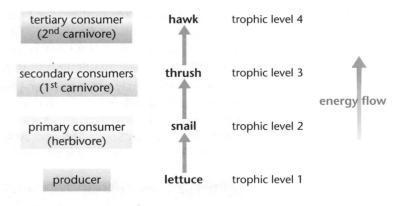

Predators and prey

There can be many examples of this type of relationship in an ecosystem. **Primary consumers** rely on the **producers**, so a flush of new vegetation may give a corresponding increase in the numbers of primary consumers. Predators which eat the primary consumers may also follow with a population increase. Each population of the ecosystem may have a **sequential effect** on other populations. Ultimately the ecosystem is in **dynamic equilibrium** and has limits as to how many of each population can survive, i.e. its **carrying capacity**.

Note that graphs are often given in predator–prey questions. A flush of spring growth is often responsible for the increase in prey. Plant biomass may not be shown on the graph! Candidates are expected to suggest this for a mark. Also remember that as prey increase, their numbers will go down when eaten by the predator. Predator numbers rise after this!

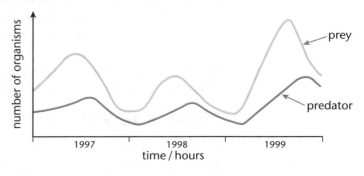

Pyramids of numbers, energy and biomass

A food chain gives limited information about feeding relationships in an area. Actual proportions of organisms in an area give more useful data. Consider this food chain from a wheat field.

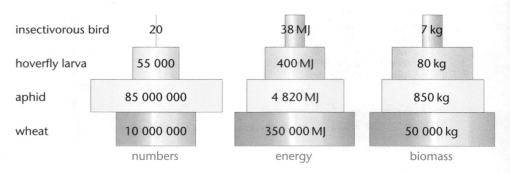

The pyramid of numbers sometimes does not give a suitable shape. In the example shown there are more aphids in the field than wheat plants. This gives the shape shown above (not a pyramid in shape!). Both the pyramids of energy and biomass are always the correct shape.

Biomass is the mass of organisms present at each stage of the food chain. The biomass of wheat would include leaves, roots and seeds. (All parts of the plant are included in this measurement.)

The organisms in the above food chain may die rather than be consumed. When this happens the decomposers use extracellular enzymes to break down any organic debris in the environment. Dead corpses, faeces and parts that are not consumed are all available for decay.

8.2 Energy transfer and agriculture

After studying this section you should be able to:

- understand a range of agricultural methods used to increase yield

How are high yields achieved in agriculture?

AQA A   M2
WJEC   M2
NICCEA   M2

The aim in agriculture is often to obtain good quality produce at maximum yield. Farmers grow crops and rear domesticated animals, like cattle, for meat and milk. Humans usually end the food chain as the top consumer. The human population is increasing constantly so efficient methods of agriculture have been developed.

Reduction of competition

Weeds reduce water, minerals and light reaching the crop plant; ultimately its rate of growth would be limited by the competing weeds. This is interspecific competition and takes place when different species need the same resources. Weeds can be removed chemically by use of a herbicide (weedkiller) or physically by an implement such as a rotavator which cuts up the weeds into tiny pieces, eventually killing them.

Intraspecific competition can also take place. This is where neighbouring plants of the same species compete for identical resources. This problem is reduced by making sure that crop plants are a suitable distance apart to achieve a maximum yield.

Use of fertilisers

It is important that crop plants have access to all the minerals they require to give a maximum yield. Farmers supply these minerals in fertilisers, usually in the form NPK (nitrogen, phosphates and potassium). By supplying them with these minerals nitrogen is available to make protein, a key substance for growth. Phosphates help the production of DNA, RNA and ATP. Potassium helps with protein synthesis and chlorophyll production. Other minerals are also needed like iron and calcium. The more a plant grows, the more its biomass increases and usually the greater is the surface area for light absorption. The amount of photosynthesis increases proportionally. If a farmer is to reach the maximum productivity of a crop, fertiliser is vital.

Increasing photosynthetic rate

As well as fertilisers helping to achieve a high productivity, other factors have a positive influence on growth:

- irrigation ensures that a plant has enough water for photosynthesis
- suitable temperature can be achieved by use of a greenhouse to give ideal conditions for the process. If a gas heater is used then the high concentration of carbon dioxide excreted can be harnessed in photosynthesis.

Pest control

If pests such as aphids or caterpillars begin to damage crops then both quality and yield are reduced. Farmers combat pests in different ways.

- Pesticides, sprayed onto crops, kill pests. Chemicals used to kill insects are

Examination questions on this topic often test knowledge of the advantages and disadvantages of each method. Chemicals may pass along food chains, and accumulate in greater quantities higher in the food chain due to the consumption of many smaller organisms, each carrying a small amount of the toxin. Biological control does not usually rely on a chemical agent so that the chemical risk is removed.

Biological control is usually much cheaper in the long term. If predators are used they go on to breed. Several repeat sprays of insecticides are needed through the growing season.

insecticides. **Contact insecticides** kill insects directly but **systemic insecticides** are absorbed into the cell sap. Any insect consuming part of the plant or sucking the sap then dies.

- **Biological control** includes a range of different methods to get rid of pests.

  (a) The most commonly used method is to use **predators** to reduce pest numbers, e.g. in greenhouses infested with whitefly (*Trialeurodes vaporariorum*) the predatory wasp (*Encarsia formosa*) is introduced. The female wasp lays eggs into the scale (larva) of the whitefly. A young wasp emerges from each larva, having used the larva as a nutrient supply. The whitefly young are killed so its population decreases. Wasps increase in numbers and remain as long as some whiteflies still remain.

  (b) **Genetic engineering** can be used. A gene has been transferred to potato plants which enables them to produce a natural insecticide. This destroys 50% of aphids that attack the plants. The amount of damage is decreased.

  (c) **Pheromones** are also used. These are compounds secreted by organisms which affect the **behaviour** within the species, e.g. the apple codling moth larva spoils the fruit by tunneling through apples. Female adult moths secrete a powerful chemical which attracts many males. This pheromone is now used in a trap. The sticky tent-shaped trap (below) shows how male moths are attracted and stick to the sides of the trap. Here they die and thousands of female moths out in the orchards are not mated and their eggs are not fertilised.

a codling moth trap

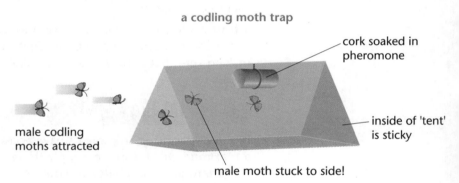

cork soaked in pheromone

male codling moths attracted

inside of 'tent' is sticky

male moth stuck to side!

  (d) **Irradiation** is used on insect pests which mate only once, e.g. the New World screw fly lays eggs in cattle (and humans!). Larvae attack the internal systems having a devastating effect. Millions of screw worm flies are bred then **irradiated**. They are subsequently released into cattle-producing regions. Irradiated males cannot produce fertile gametes. Any male mating with a female from the cattle fields results in unfertilised eggs so the population decreases. Cattle productivity is maintained.

millions of irradiated screw flies released from a plane

8.3 Nutrient cycles

After studying this section you should be able to:

- *recall how carbon and nitrogen are recycled*

The nitrogen cycle

EDEXCEL	M3
OCR	M1
WJEC	M1
NICCEA	M2

Nitrogen is found in every amino acid, protein, DNA and RNA. It is an essential element! Most organisms are unable to use atmospheric nitrogen directly so the nitrogen cycle is very important.

There are three parts of the nitrogen cycle which are regularly examined:
- nitrogen fixation in leguminous plants
- nitrification
- denitrification.

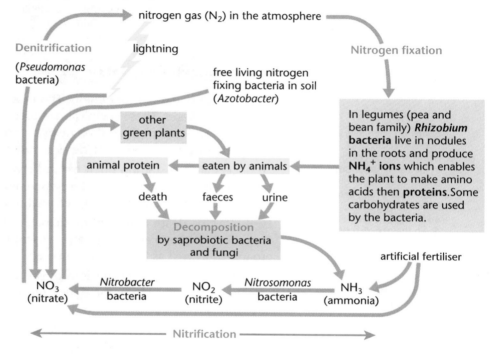

The association of *Rhizobium* bacteria with legume plants give advantages to both organisms. This relationship is known as **mutualism**.

Saprobiotic bacteria and fungi secrete extracellular enzymes. They obtain nutrients in this way.

The biochemical route from ammonia to nitrate is **nitrification**. This is helped by ploughing which allows air into the soil. Nitrifying bacteria are aerobic. Draining also helps.

Some important points

- **Nitrogen gas** from the atmosphere is used by *Rhizobium* **bacteria**. These bacteria, living in nodules of legume plants, convert nitrogen gas into ammonia (NH_3) then into amine($-NH_2$) compounds. The plants transport the amines from the nodules and make amino acids then proteins. *Rhizobium* bacteria gain carbohydrates from the plant, therefore each organism benefits.

- Plants support food webs, throughout which excretion, production of faeces and death take place. These resources are of considerable benefit to the ecosystem, but first decompostion by saprobiotic bacteria takes place, a waste product of this process is ammonia.

- Ammonia is needed by *Nitrosomonas* bacteria for a special type of nutrition (chemo-autotrophic). As a result another waste product, nitrite (NO_2) is formed.

- Nitrite is needed by *Nitrobacter* bacteria, again for chemo-autotrophic nutrition. The waste product from this process is nitrate, vital for plant growth. Plants absorb large quantities of nitrates via their roots.

- Nitrogen gas is returned to the atmosphere by denitrifying bacteria such as *Azotobacter*. Some nitrate is converted back to nitrogen gas by these bacteria. The cycle is complete!

The carbon cycle

EDEXCEL M3
OCR M1
NICCEA M2

Carbon is the key element in all organisms. The source of this carbon is atmospheric carbon dioxide which proportionally is 0.03% of the volume of the air. Most organisms cannot use carbon dioxide directly so the carbon cycle is very important.

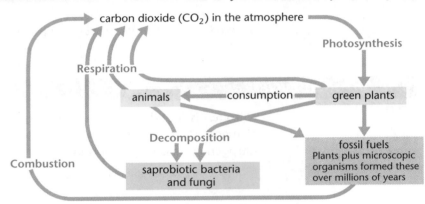

Some important points

- Producers carry out photosynthesis. This process incorporates the carbon dioxide into carbohydrates. These chemicals are used as a starting point to make lipids and proteins. Some of the carbon helps to form structures in the producers and some is released as carbon dioxide as a waste product of respiration.

- Producers are the starting point of food chains. After the plants are eaten by primary consumers carbon can be passed along to subsequent consumers. It can be incorporated into tissues, respired or excreted.

- Even after the death of a plant or animal, carbon dioxide can still be released. Saprobiotic bacteria and fungi respire using the organic chemicals in dead organisms as well as faeces and urine, etc.

- Compression of organisms millions of years ago resulted in the formation of fossil fuels. Combustion of these fuels releases carbon dioxide back into the atmosphere.

- The return of carbon dioxide to the air completes the cycle!

Other elements are also recycled. The decomposers have a major role in maintaining the availability of vital chemicals.

> Examiners often give a question about what happens to the energy in the chemicals of dead organisms or organic waste. Many candidates correctly state that microorganisms rot down the materials but then go on to state that energy goes into the ground. Big mistake! Energy is released by the respiration of the decomposers to support their life.

8.4 Colonisation and succession

After studying this section you should be able to:

- *understand how colonisation is followed by changes*
- *understand how colonisation and succession lead to a climax community*

LEARNING SUMMARY

How decolonisation and succession take place

WJEC M2
NICCEA M2

Any area which has never been inhabited by any organisms may be available for **primary succession**. Such areas could be a garden pond filled with tap water, lava having erupted from a volcano, or even a concrete tile on a roof. The latter may become colonised by lichens.

Occasionally an ecosystem may be destroyed, e.g. fire destroying a woodland. This allows **secondary succession** to begin, and signals the reintroduction of plant and animal species to the area.

Colonisation and succession also take place in water. Even an artificial garden pond would be colonised by organisms naturally. Aquatic algae would arrive on birds' feet.

The process of succession can take place as follows.

- **Pioneer species (primary colonisers)** begin to exploit a 'new' habitat. Mosses may successfully grow on newly exposed heathland soil. These are the primary colonisers which have adaptations to this environment. Fast germination of spores and the ability to grow in waterlogged and acid conditions, aid rapid colonisation. These plants may support a specific food web. In time, as organic matter drops from these herbaceous colonisers it is decomposed (see nitrogen cycle page 115), nutrients are added to the soil and acidity increases. In time the changes caused by the primary colonisers cause the habitat to be unsuitable.

- Conditions unsuitable to primary colonisers may be ideal for other organisms. In early heathland, mosses are replaced by heathers which can thrive in acid and xerophytic (desiccating) conditions. This is succession, where one community of organisms is replaced with another. In this example the secondary colonisers have replaced the primary colonisers; this is known as seral stage 1 in the succession process. Again, a different food web is supported by the secondary colonisers.

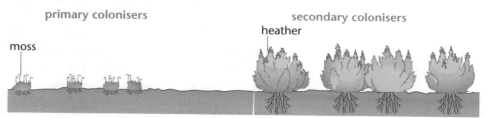

primary colonisers

moss

secondary colonisers

heather

- At every seral stage there are changes in the environment. The second seral stage takes place as the tertiary colonisers replace the previous organisms. In heathland, the new conditions would favour shrubs such as gorse and bilberry plus associated animals.

- The shrubs are replaced in time with birch woodland, the third seral stage. Eventually acidic build-up leads to the destruction of the dominant plant species.

- Finally conditions become suitable for a dominant plant species, the oak. Tree saplings quickly become established. Beneath the oak trees, grasses, ferns, holly and bluebells can grow in harmony. This final stage is stable and can continue for hundreds of years. This is the climax community. Associated animals survive and thrive alongside these plant resources. Insects such as gall wasps exploit the oak and dormice eat the wasp larvae. Jays are birds which eat some acorns but spread others which they store and forget. The acorns germinate; the woodland spreads.

climax community

oak woodland

In Britain, an excellent example of a climax community is Sherwood Forest where the 'Major Oak' has stood for 400 years. Agricultural areas grow crops efficiently by **deflecting succession**. Plants and animals in their natural habitat are 'more than a match' for domesticated crops. Herbicides and pesticides are used to stop the invaders!

8.5 Effects of human activities on the environment

After studying this section you should be able to:

- *understand some causes and effects of pollution*
- *understand the effects of deforestation*
- *understand the problems of over-fishing*

LEARNING SUMMARY

How human activities affect the environment

AQA A   M2
EDEXCEL   M3
WJEC   M2
NICCEA   M2

Activities carried out by the human population to supply food, power, and industrial needs have a considerable effect on the environment. These effects include atmospheric and water pollution, and destroying habitats and communities.

What is the greenhouse effect?

This is caused by specific gases which form a thin layer around the atmosphere. These gases include water vapour, carbon dioxide, methane, ozone, and nitrogen oxides and CFCs. CFCs (CCl_2F_2, CCl_3F) have a greenhouse factor of 25 000 based on the same amount of carbon dioxide at a factor of 1.0. The quantity of the greenhouse factor gas needs to be considered to work out the overall greenhouse effect, e.g.

> carbon dioxide is 0.035% of the troposphere × greenhouse factor value 1= 0.035
>
> CCl_2F_2 is 4.8×10^{-8} % of the troposphere × greenhouse factor value 25 000 = 0.012
>
> Water vapour is 1% of the troposphere × greenhouse factor 0.1 = 0.1

KEY POINT

It is clear that carbon dioxide has the greatest overall greenhouse effect!

- The greenhouse gases allow short wavelength radiation from the sun to reach the Earth's surface.
- Some of the infra-red radiation fails to pass back through the greenhouse layer resulting in global warming.
- Polar ice caps may melt causing the sea to rise and subsequent reduction of land mass. Some aquatic populations could increase and some terrestrial populations decrease.
- Climatic changes are expected, so rainfall changes and heat increases will have significant effects.

Examiner's tip.
Do not mix up the greenhouse effect with the 'hole in the ozone layer' – that is different! The ozone layer around the Earth absorbs some ultra-violet radiation from the sun. If a lot of ultra-violet radiation reaches the Earth's surface then many people succumb to skin cancer. CFCs cause a hole to form in the ozone layer. Not using these chemicals is the answer to this problem.

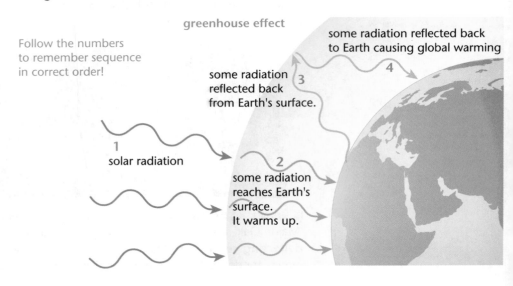

greenhouse effect

Follow the numbers to remember sequence in correct order!

1 solar radiation

2 some radiation reaches Earth's surface. It warms up.

3 some radiation reflected back from Earth's surface.

4 some radiation reflected back to Earth causing global warming

Deforestation

In countries such as Brazil, forests have been burned down. Large quantities of carbon dioxide and water vapour released into the atmosphere contribute to the greenhouse effect. However, the long-term effects are highly significant. Habitats and complete food webs are lost. Biodiversity is decreased so that many less species are represented on the land left after deforestation. The canopy of a forest intercepts and holds rain water, so too much rain does not reach the ground and cause flooding. Instead, much water evaporates back into the atmosphere. Without the forest, flooding is a danger, and without the tree roots, soil erosion takes place. If the deforestation was to make way for agriculture then there are major problems. Top soil is lost and nutrients leach into the ground and in the long term agriculture fails.

Water pollution

'Run off', containing fertilisers, enters rivers from fields. Similarly, sewage also pollutes rivers. The diagrams below show a river before and after sewage entry.

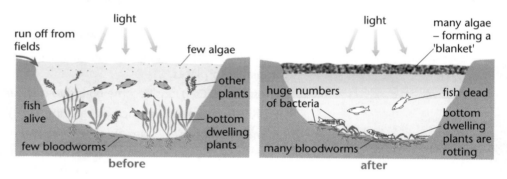

The polluting effect of fertilisers and sewage are caused by the constituent ions such as nitrates and phosphates. They result in **eutrophication**.

- **Nitrates and phosphates** enter the river and are absorbed by plants. This promotes plant growth.

- **Algae** float near the water surface and their population increases dramatically. A 'blanket' of algae soon covers the surface.

- Bottom dwelling plants do not share the same advantage. Initially the ions promote growth but **surface algae block the sunlight**. Plants beneath the algae die.

- **Bacteria** and other decomposers begin to **break down** the **dead plants** and some short lived algae. The **bacterial population increases** and proportionally takes more oxygen from the river water.

- **Fish die** as the oxygen content becomes much too low. Rotting dead fish contribute to even lower oxygen levels, again aerobic bacteria are responsible.

- Often there is an **increase in bloodworms** (tubifex). These are mud dwellers and possess a protein similar to haemoglobin which helps them to take in enough oxygen for survival, even at low concentration. Without fish to eat them numbers of bloodworm increase even more.

Examiner's tip.

The sequence of stages of eutrophication are important. Take care to learn the correct sequence. Do not miss out any stage. Some candidates merely state nitrates enter the river and fish die. Give the detail and score more marks!

How can the water pollution be measured?

There are many ways to measure both pollutants and their effects. Populations of algae, bloodworms or fish can be estimated. A key measurement is **biological oxygen demand (BOD)**. This is the amount of oxygen taken up by a sample of water at 20°C over 5 days. Clean water takes up much less oxygen than that polluted with organic material. Aerobic bacteria take up a large proportion of this oxygen. A river, heavily polluted with organic matter, has a very high BOD.

Indicator species

The presence or absence of a species in the river can be used as a sign of pollution. Mayfly larvae can only tolerate well oxygenated water. Bloodworms are only found in large numbers in water heavily polluted with organic matter. As the river flows downstream, organisms change the organic matter and eventually the oxygen content increases. A large population of mayfly larvae found in water downstream suggests that there is:

* a low BOD
* organic material further upstream which has been changed by bacteria so the water is no longer polluted.

Acid rain

> There are other acid gases apart from sulphur dioxide, e.g. nitrogen oxides.

This is caused by the combustion of fossil fuels, e.g. coal. This releases a number of acidic gases which dissolve in rain water. One of the most significant of these gases is sulphur dioxide.

$$H_2O \ + \ SO_2 \ = \ H_2SO_3$$

| rain water | sulphur dioxide | sulphurous acid |

Rain of low pH can have a devastating effect on the organisms of an ecosystem.

* Low pH results in many mineral ions being less soluble and consequently less available to plants.
* Phosphate (PO_4^-) ions become bound to clay particles and are unavailable to plants.
* Positively charged ions such as calcium (Ca^{2+}) are more easily leached.
* Aluminium ions (Al^{3+}) are an exception and may accumulate to a toxic level.
* Plants may be defoliated and die. This has a sequential effect on all consumers which rely directly or indirectly on the plants in food webs.
* In lakes the low pH destroys organisms, e.g. fish, often as a result of Al^{3+} build up.

> Remember that 'oligotrophic' is an opposite term to eutrophic.

* Levels of other minerals are normally low and cannot sustain much plant growth. A lake is said to be oligotrophic in this condition.

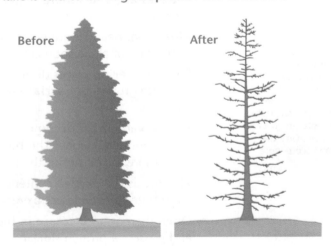

Before   After

> Examiners often give questions about the acid rain problem. Higher grade students note the consequences of trees dying. Complete food webs can be destroyed!

What is the answer to pollution?

People need to make personal choices. Do they support the use of products which have involved pollution? Do they support the use of legislation to prevent pollution?

> **KEY POINT**
>
> In Europe currently there is legislation to ensure that each country complies with targets to reduce both water and air pollution. It is vital that we make the correct decisions if biodiversity is to be maintained.

Progress check

(a) Explain how acid rain is formed.

(b) Suggest the effects that acid rain may have on a woodland ecosystem.

(c) A student states that all power stations contribute to acid rain. Is this true or false? Give a reason for your answer.

(a) Acid rain is caused by the burning of fossil fuels, e.g. oil. Acidic gases dissolve in rain water, e.g. sulphur dioxide.

$$H_2O + SO_2 = H_2SO_3$$
water   sulphur   sulphurous
        dioxide   acid
rain

Cloud water vapour condenses to produce acid rain.

(b) Plants become defoliated and die; some minerals cannot be absorbed; phosphate ions become bound to clay particles and are unavailable to plants; positively charged ions such as calcium are leached; aluminium ions accumulate to a toxic level; there is a sequential effect on all consumers which rely directly or indirectly on the plants in food webs.

(c) False. Only power stations which are fuelled by fossil fuels emit acid gases. Nuclear power stations just emit water vapour from the cooling towers.

Remember that over-fishing may have a sequential effect on other organisms of an ecosystem. Some organisms may increase in numbers because fish are no longer eating them, e.g. zooplankton. Other organisms may decrease in numbers, e.g. phytoplankton, because fewer zooplankton are consuming them.

Problems in over-fishing

The human population needs food, but as the World population has increased more food is required. Just as agriculture has improved productivity, the fishing industry has improved techniques of catching fish.

- Echo-sounding equipment is used to locate precisely fish populations.
- Giant trawler nets drag across the sea floor to catch demersal (bottom dwelling) fish.
- Giant drift nets are suspended in the sea so that unsuspecting pelagic (surface dwelling) fish bump into the nets and become trapped by their gill flaps.
- Nets often have a small mesh so that both small and large fish are caught.

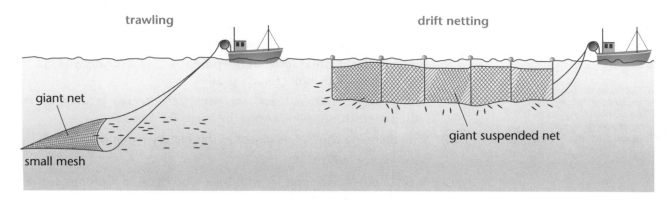

Fish are returned to the sea if the numbers caught are beyond the fishing quota or if the fish are too small. Perhaps fish farms are the answer.

Techniques are so efficient that breeding stocks have decreased dramatically. Many small fish never reach maturity. Legislation is the answer to increase mesh size, enforce quotas and exclusion zones. By international agreement, breeding stocks may be allowed to become re-established.

Sample question and model answer

Some questions give information that you can use to deduce the answers (a) (i) is one of them!

(a) The sequence below shows how nitrate can be produced from a supply of oak leaves.

decomposers   Nitrosomonas   Nitrobacter
              bacteria   bacteria

dead oak -----> NH_3 -----> NO_2 ----- > NO_3
leaves   (ammonia)   (nitrite)   (nitrate)

This question targets the nitrogen cycle. Be ready to answer questions about *any part* of the cycle. Pure recall of the cycle is not enough! You need to apply your knowledge.

(i) Suggest the consequences of death of the Nitrosomonas bacteria. [4]

build up of ammonia; build up of dead leaves; death of Nitrobacter bacteria; no nitrite/no nitrate

(ii) Name the process by which bacteria produce nitrate from ammonia. [1]

nitrification

(iii) Name **two** populations of organisms not shown in the sequence which would be harmed by a lack of nitrate. [2]

denitrifying bacteria or Pseudomonas; plants or producers

(iv) Which organisms fix atmospheric nitrogen on the nodules of bean plants? [1]

(Rhizobium) bacteria

(b) Apart from adding fertiliser or irrigating a crop, how can a farmer make sure of producing a high yield ? [4]

In questions targeting yield in agriculture make sure that you have clarified all terms, e.g. do not confuse herbicide with pesticide.

make sure that the plants are the correct distance apart; use of pesticide or use of insecticide; use of herbicide or fungicide; use of biological control or named biological control; use of variety produced by selective breeding [any four]

(c) (i) A farmer rears pigs by a factory-farming method. Pigs are kept indoors 24 hours per day in warm, confined cubicles.

How can this method result in the production of a greater yield of pork than from animals reared outside? [3]

Note that the reverse explanation for pigs reared outside can be given, i.e. more energy used for movement, etc.

less energy is released for movement; less energy is used to maintain body temperature; more energy is used for biomass

(ii) Why do many consumers object to this factory farming method? [1]

cruel or not ethical

(iii) Pigs are often given copper with their food because it promotes their growth. Suggest **one** disadvantage of using this method. [2]

it may contaminate the pork; people eat the pork and may be harmed

Practice examination questions

1 The number of species of grass and the number of leguminous plants growing in two fields was measured over a 10-year period. Field A was given nitrogenous fertiliser each year, but field B was given none. The results are shown in the graphs.

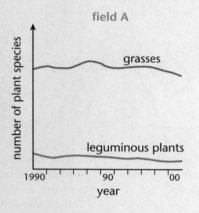

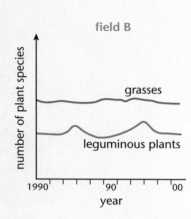

(a) (i) Suggest why there were fewer leguminous plant species in field A. [2]

    (ii) Suggest why there were more leguminous plant species in field B. [2]

(b) After the main investigation no fertiliser at all was used in either field. Cattle were allowed to graze in both fields. At the end of five years the number of legume species in each field had decreased. Suggest why the number of legume plants decreased. [1]

2 The following article appeared in a newspaper.

> ### ATLANTIC COD IN DANGER OF EXTINCTION!
> The cod stocks of the Atlantic Ocean are in urgent need of protection. Deep-water trawling is so efficient that in the last 10 years numbers have fallen drastically. A survey found that the average length of the fish has decreased by 10 cm
> 5   since 1990. This will have a harmful effect on the population and could ultimately lead to extinction. One of the problems is improved technology which locates complete shoals of the fish. Trawl nets are dragged along the sea floor stirring up sediment and killing many delicate
> 10   invertebrates. This will affect biodiversity. So many countries fish the waters that a political agreement must be found.

Extract from 'Action Ecology' 2000

(a) Suggest **two** ways in which trawling is considered 'efficient'. (line 2) [2]

(b) Why was the decrease in the average length of the cod considered to have a 'harmful effect' on the population of the species? (line 5) [2]

(c) How may trawling result in reduced biodiversity? (line 10) [3]

(d) Suggest how 'political agreement' may help to increase breeding stocks. (line 11) [2]

Practice examination questions (continued)

3 The diagrams show stages in the development of a garden pond over a 10-year period.

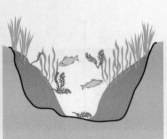
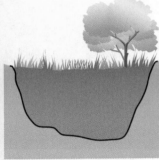

A hole was dug, lined with butyl and new plants were placed in the pond.	Marginal plants grow, spread and die down in the winter. As they rot sediment falls to the bottom of the pond.	After a number of years the pond has completely covered over.
1990	1995	2000

(a) In 1990 irises, oxygenating pondweed and a water lily were planted in the pond. Algae were not planted but arrived in the pond in some other way.

   (i) What term describes an organism that grows in a new habitat that previously supported no life? [1]

   (ii) After a time the algae produced a thick 'carpet' of growth on the surface of the pond. Explain the effect this may have on organisms under the water. [5]

(b) Describe the stages which took place to produce the stable grassland after 10 years. [2]

4 (a) Explain how the increase in mesh size of trawling nets can help increase fish stocks in the sea. [2]

(b) Apart from the increase of mesh size, how can governments increase fish stocks in the sea? [1]

(c) A herbivorous species of fish is part of a food chain. If the fish species increased in number, suggest **two** effects this may have on the other organisms of the food chain. [2]

5 The following two gases help cause the greenhouse effect.

	Greenhouse effect factor	*Relative amount in troposphere*
Water vapour	0.1	1%
CFC'S	25 000	4.8×10^{-8}

(a) Work out which gas has the greatest influence on the greenhouse effect. [2]

(b) Suggest **one** reason for the greenhouse effect resulting in:

   (i) an increase in the population of a species [1]

   (ii) a decrease in the population of a species. [1]

9 Human health and disease

- *Health and lifestyle*
- *Disease*
- *Immunity*

9.1 Health and lifestyle

After studying this section you should be able to:

- *define health*
- *understand the features of good health*
- *understand how good health can be maintained through diet and exercise*
- *describe the specific effects of smoking tobacco*

How can we achieve good health?

AQA A ► M1
OCR ► M2

Good health is not just an absence of disease or infirmity. It is the physical, mental and social well-being of a person. The development of a healthy person begins in the uterus. It is important that the mother supplies the fetus with suitable nutrients for development, e.g. amino acids for proteins essential for healthy growth. The mother's diet is important for both her and the fetus.

Following birth, the emotional and social development are equally as important as physical development.

If a person is healthy then they may expect the following:

- an absence of disease
- an absence of pain
- to be fit and have good muscle tone
- an absence of stress
- to get along with other people in society
- to have a long life expectancy.

The importance of diet

The human diet is vital to good health. A newly born baby needs to feed on mother's first milk (**colostrum**) which is rich in antibodies. This gives immunity against some diseases. It is important that each of the following food classes are included in a person's diet, in a suitable proportion.

The balanced diet

Examiner's tip
It is likely that you will be supplied with data about dietary constituents. Be ready to analyse the data and apply the principles of a balanced diet. Try to remember the main function of each substance. Analyse the dietary reference values for food energy and nutrients in the UK. The values indicate amounts of individual food components required per day and those which should not be exceeded.

- **Carbohydrates** – sugars and starch supply metabolic energy; cellulose (dietary fibre) stimulates peristalsis so that constipation is prevented.
  Source – potato and bread

- **Proteins** – supply metabolic energy and are needed in growth and repair. **All enzymes are proteins.** Very important!
  Source – meat and nuts

- **Fats and oils** – supply metabolic energy and are needed in cell membrane formation, as they help to make phospholipids.
  Source – butter and cooking oil

- **Vitamins** – organic substances needed in minute quantities to maintain health, e.g. vitamin A. This is essential to make the pigment in the rods of the retina.
  Source (vitamin A) – butter and carrots

- **Minerals** – inorganic ions needed for a number of important roles in the body, e.g. iron. This is essential for the production of haemoglobin and so is vital for oxygen transport.
  Source (iron) – red meat, spinach
- **Water** – makes up over 50% of the content of blood plasma. It is needed for many functions including as a solvent and cooling the body down.

If a person does not eat enough of any one constituent of their diet then there is a deficiency disease, e.g. a protein deficiency causes kwashiorkor. If a person eats too much carbohydrate and fats then obesity and cardiovascular problems can result. A balanced diet is vital!

> Remember that carbohydrates, proteins, lipids, vitamins, minerals, water and dietary fibre are all essential.

Essential amino acids

All the parts of a balanced diet are vital if good health is to be maintained. Proteins supply amino acids which can be used as an energy source or to build different proteins. There are 20 different amino acids used to make important proteins. Children need 10 essential amino acids but adults need just 8.

Essential amino acids must be supplied in the diet and cannot be made in the body. A person who eats foods with all the essential amino acids is able to make the others.

Daily energy requirement

Eating, then respiring carbohydrates, proteins, fats and oils, supplies the energy needed for good health. Energy content of food is usually measured in kilojoules (kJ). A diet rich in carbohydrates and/or fats and oils which exceeds daily requirements results in obesity. Large quantities of fat are stored around the body resulting in cardiovascular problems. If the kilojoule intake is regularly less than the daily requirement then a condition known as anorexia nervosa can develop. People with this condition are unable to eat enough food and they lose body mass.

> Exercise does much more than reduce weight! It develops good muscle tone, keeps joints supple and even mental health is positively affected.

What are the energy needs of different people?

Energy needs are determined by age, gender and activity. Additionally, if a woman is pregnant or lactating, extra food is needed. The table below shows typical daily energy requirements.

Person	Energy used in one day (kilojoules)			
baby (0–3 months)	2 400			
infant (1 year)	4 300			
child (8 years)	8 900			
teenager (15 years)		Male 12 600	Female	9 600
adult (office work)		Male 11 600	Female	9 500
adult (heavy work)		Male 16 600	Female	12 600
pregnant woman	10 500			
lactating mother	11 400			

When a person is sleeping their respiratory rate is at a minimum, so the amount of energy released from food is also at a minimum. A person may be involved in a very physical activity like building work or mountain biking; on these occasions energy requirement is large. Exercise is important for health. It can prevent excessive weight increase since more kilojoules of energy are released by increased respiratory activity.

If the energy component and the other components of the diet are less than the recommended daily values, malnutrition takes place. This resembles multiple deficiency diseases. Muscle wastage follows and the individual is in great danger.

Saturated fat and coronary heart disease

The diet is important if we are to maintain health. People can make choices. They can eat a balanced diet and avoid problems caused by eating food components in the wrong proportions.

An example of part of the diet which should be eaten in small quantities is saturated fat. Saturated fats (see lipid structure page 24) are found in large quantities in animal tissues.

> **KEY POINT**
>
> Eating red meat such as beef, fatty pork chops, sausages and dairy products can result in health problems. The way that the food is cooked can also result in health problems. Frying food in animal fat adds to the danger! Many people consume the above foods but in smaller quantities. Dietary balance is important.

Atherosclerosis

This is a major health problem caused by eating saturated fats. This circulatory disease may develop as follows:

- yellow fatty streaks develop under the lining of the endothelium on the inside of an artery
- the streaks develop into a fatty lump called an atheroma
- the atheroma is made from cholesterol (taken up in the diet as well as being made in the liver)
- dense fibrous tissue develops as the atheroma grows
- the endothelial lining can split, allowing blood to contact the fibrous atheroma
- the damage may lead to a blood clot and an artery can be blocked.

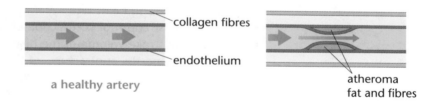

collagen fibres
endothelium

a healthy artery

atheroma
fat and fibres

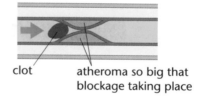

clot   atheroma so big that blockage taking place

Remember that the clotting of blood can occur for other reasons. There may be damage at other positions around the body. Blockage of this type is **thrombosis**.

Increasing constriction of an artery caused by atherosclerosis and blood clots reduces blood flow and increases blood pressure. If the artery wall is considerably weakened then a bulge in the side appears, just like a weakened inner tube on a cycle tyre. There is a danger of bursting and the structure is known as an aneurysm.

It is possible for a blood clot formed at an atheroma to break away from its original position. It may completely block a smaller vessel, this is known as an embolism.

If the artery which supplies the heart (coronary artery) is partially blocked, then there is a reduction in oxygen and nutrient supply to the heart itself. This causes angina, the main symptom being sharp chest pains. If total blockage occurs then myocardial infarction (heart attack) takes place.

Other aspects of lifestyle influence the condition of the cardiovascular system. A combination of factors are responsible for our health.

Progress check

1 (a) The diet of an infant must contain 10 **essential** amino acids to help maintain health. What is an essential amino acid?
 (b) How many essential amino acids are needed by an adult person?

2 (a) Name a specific substance in food which can result in atherosclerosis.
 (b) Describe and explain the structural changes which take place in a blood vessel as atherosclerosis develops.
 (c) How can the damage caused by an atheroma result in a heart attack?

1 (a) an amino acid which must be supplied in the diet. Non-essential amino acids can be made within the body. (b) 8
2 (a) saturated fat
 (b) yellow fatty streaks develop under the cells lining the inside of a blood vessel, the streaks develop into a lump known as an atheroma, the atheroma is made of cholesterol, dense fibrous tissue develops and the lining of the vessel can split.
 (c) a blood clot forms which can block the blood vessel completely. Prevention of oxygen supply to the heart results in myocardial infarction (heart attack).

Effects of lifestyle

AQA A ▶ M1
OCR ▶ M2

Statistically people have a greater chance of living longer if they:

- do not smoke
- do not drink alcohol excessively
- consume a low amount of salt
- consume a low amount of saturated fat in their diet
- are not stressed most of the time
- exercise regularly.

Exercise has a **protective effect** on the **heart** and **circulation**. Activities such as jogging, walking, swimming and cycling can build up the person's endurance.

It is not enough to do a minor amount of exercise, infrequently. The **intensity**, **frequency** and **duration** of the exercise are all important if a programme is to be effective. Frequent exercise:

- reduces the resting heart rate
- increases the strength of contraction of the heart muscle
- increases the stroke volume of the heart (the volume of blood which is propelled during the contractions of the ventricles)
- aids mobility and subtlety of the body
- increases the rate of recovery after a strenuous activity so that heart rate and breathing rates return to resting levels more quickly.

Recommended amount of exercise per week:

- **Intensity** – should allow your heart to beat at a minimum of 60% of your maximum heart rate and increase as you become more fit
- **Frequency** – around three times weekly
- **Duration** – about 30–60 minutes per session.

The diagrams below show the typical effect of training on a person's heart.

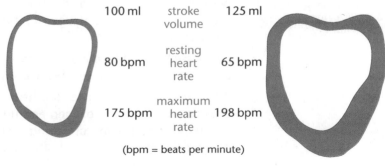

	heart before training		heart after training
stroke volume	100 ml		125 ml
resting heart rate	80 bpm		65 bpm
maximum heart rate	175 bpm		198 bpm

(bpm = beats per minute)

What are the dangers of smoking tobacco?

OCR ▶ M2

Each person has another choice to make, to smoke or not to smoke. The government health warning on every cigarette packet informs of health dangers but many young people go ahead and ignore the information.

Effects of tobacco smoking

- Nicotine is the active component in tobacco which addicts people to the habit.

- Tars coat the alveoli which slows down exchange of carbon dioxide and oxygen. If less oxygen is absorbed then the smoker will be less active than their true potential.

- Cilia lining bronchial tubes are coated then destroyed, this reduces the efficiency in getting rid of pollutants which enter the lungs. These pollutants include the cigarette chemicals themselves.

- Carbon monoxide from the cigarette gases combines with haemoglobin of red blood cells rather than oxygen. This reduces oxygen transport and the smoker becomes less active than their potential. Ultimately it may lead to heart disease.

- The bronchi and bronchioles become inflamed, a condition known as bronchitis. This causes irritating fluid in the lungs, coughing and increased risk of heart disease. A number of bronchitis sufferers die each year.

- The walls of the alveoli break down reducing the surface area for gaseous exchange. Less oxygen can be absorbed by the lungs, leaving the emphysema sufferer extremely breathless. They increase their breathing rate to compensate but still cannot take in enough oxygen for a healthy life. A chronic emphysema sufferer needs an oxygen cylinder to prolong their life. Death is a regular conclusion, especially when combined with other symptoms.

- Blood vessel elasticity is reduced so that serious damage may occur. Ultimately a heart attack can follow.

- The carcinogens (cancer-causing chemicals) of the tobacco can result in lung cancer. Malignant growths in the lungs develop uncontrollably and cancers may spread to other parts of the body. Death often follows. Smokers have a greater risk of developing other cancers than non-smokers, e.g. more smokers develop cervical cancer.

Even non-smokers can develop any of the above symptoms, but the probability of developing them is increased by smoking. Being in a smoky atmosphere each day, such as a non-smoker working in a pub, also increases the chances.

> Learn the characteristics of each disease carefully. There are so many consequences of smoking that you may well mix them up.

The role of statistics

The government health warning on cigarette packets informs people of the risks of smoking. The WHO (World Health Organisation), governments and local authorities have collected statistics on many diseases over the years. These are used in education packs and posters to warn of risk factors. People can take precautions and use the information to avoid health dangers and take advantage of vaccination programmes.

Where education is not successful then related diseases follow. This acts as a drain on the National Health Service. Many operations which would have been unnecessary are performed to save people's lives, e.g. where coronary blood vessels are dangerously diseased a by-pass operation is the answer.

The ideal situation is that education is successful, but realistically the aim is to balance prevention and cure.

9.2 Disease

After studying this section you should be able to:

- *define disease*
- *recall the causes, symptoms and control of a range of diseases including cholera, tuberculosis, malaria and AIDS*

LEARNING SUMMARY

What is a disease?

 OCR · M2

A disease is a disorder of a tissue, organ or system of an organism. As a result of a disorder, symptoms are evident. Such symptoms could be the failure to produce a particular digestive enzyme, or a growth of cells in the wrong place. Normal bodily processes may be disrupted, e.g. efficient oxygen transport is impeded by the malarial parasite, *Plasmodium*.

Different types of disease

Infectious disease by pathogens

Pathogens attack an organism and can be passed from one person to another. Many pathogens are spread by a vector which carries it from one organism to another without being affected itself by the disease. Pathogens include bacteria, viruses, protozoa, fungi, parasites and worms.

> Most exam candidates recall that pathogens are responsible for disease. However, there are more causes of disease! If a question asks for different types of disease then giving a range of pathogens will not score many marks. Give genetic diseases, etc.

Genetic diseases

These can be passed from parent to offspring. Also known as congenital diseases, they include haemophilia and cystic fibrosis.

Dietary related diseases

These are caused from the foods that we eat. Too much or too little food may cause disorders, e.g. obesity or anorexia nervosa. Lack of vitamin D causes the bone disease rickets, the symptoms of which are soft weak bones which bend under the body weight (see page 126 deficiency diseases).

Environmentally caused diseases

Some aspects of the environment disrupt bodily processes, e.g. as a result of nuclear radiation leakage, cancer may develop.

> An auto-immune disease may be environmentally caused, e.g. the form of leukaemia where phagocytes destroy a person's red blood cells may be caused by radiation leakage.

Auto-immune disease

The body in some way attacks its own cells so that processes fail to function effectively.

How are infectious diseases transmitted?

The pathogens which cause infectious diseases are transmitted in a range of ways.

- Direct contact – sexual intercourse enables the transmission of syphillis bacteria; a person's foot which touches a damp floor at the swimming baths can transfer the Athlete's foot fungus.
- Droplet infection – a sneeze propels tiny droplets of nasal mucus carrying viruses such as those causing influenza.
- Via a vector – if a person with typhoid bacteria in the gut handles food the bacteria can be passed to a susceptible person.
- Via food or water – chicken meat kept in warm conditions encourages the reproduction of *Salmonella* bacteria which are transferred to the human consumer, who has food poisoning as a result.

> Be prepared to answer questions about diseases not on your syllabus. The examiners will give data and other information which you will need to interpret. Use your knowledge of the principles of disease transmission, infection, symptoms and cure.

- Via blood transfusion – as a result of receiving blood a person can contract AIDS.

Some infectious diseases have serious consequences to human life. The incidence of infectious diseases may vary according to the climate of the country, the presence of vectors, the social behaviour of people and other factors.

The infectious diseases in an area may be classified by using the following terms:

(a) endemic, which means that a disease or its vector is invariably found in an area

(b) epidemic, which means that there is an outbreak of a disease attacking many people in an area

(c) pandemic, which means that a there is an outbreak of a disease over a very large area, e.g. the size of a continent.

Disease file – cholera

OCR   M2

Cause of disease

Vibrio cholerae (bacteria) in the faeces or vomit of a human sufferer or human carrier which contaminate water supplies.

Transmission of microorganism

Contaminated water spreads the bacteria. Poor sanitary behaviour of people who are carriers and those who have contracted the disease are responsible. Faeces enters rivers which may be used for bathing, drinking, or irrigation. The bacteria survive outside the human body for around 24 hours. They can also contaminate vegetables and can be passed to a person in this way.

Outline of the course of the disease and symptoms

The bacteria reach the intestines where they breed. They secrete a toxin which stimulates adenyl cyclase in epithelial cells. This enzyme causes much fluid to be secreted into the intestine, giving severe diarrhoea. Death is a regular consequence, due to dehydration, but some people do recover.

Prevention

Education about cleanliness and sewage treatment. Good sanitation is vital. Suitable treatment of water to be consumed by people, e.g. chlorination which kills the bacteria. Use of disinfectant also kills the bacteria. Early identification of an outbreak followed by control.

Cure

Tetracycline antibiotics kill organisms in the bowel. Immunisation is not very effective. It will help some individuals but not stop them from being carriers, so epidemics are still likely.

Disease file – tuberculosis

OCR   M2

Cause of disease

Mycobacterium tuberculosis (bacterium) via droplet infection.

Transmission of microorganism

Coughs and sneezes of sufferers spread tiny droplets of moisture containing the pathogenic bacteria. People then inhale these droplets and may contract the disease.

Outline of the course of the disease and symptoms

The initial attack takes place in the lungs. The alveoli surfaces and capillaries are vulnerable and lesions occur. Some epithelial tissues begin to grow in number but these cannot carry out gaseous exchange. Inflammation occurs which stimulates painful coughing. Intense coughing takes place which can cause bleeding. There is much weight loss. Weak groups of people, like the elderly, or someone underweight are more prone to the disease.

Prevention

Mycobacterium bovis causes tuberculosis in cattle. It can be passed to humans via milk. It causes an intestinal complaint in humans. It is important that cows are kept free of *M. bovis* by antibiotics.

The BCG vaccination is the injection of a weakened form of this microbe. This vaccination stimulates antibodies which are effective against both *M. tuberculosis* and *M. bovis*.

Mass screening using **X-rays** can identify 'shadows' in those people with scar tissue in the lungs.

Sputum testing identifies the presence of the bacteria in sufferers. Sufferers can be treated with antibiotics. Once cured they cannot pass on the pathogen so an epidemic may be prevented.

Skin testing is used. Antigens from dead *Mycobacteria* are injected just beneath the skin. If a person has been previously exposed to the organism then the skin swells which shows that they already have resistance, i.e. they have antibodies already. Anyone whose skin does not swell up is given the **BCG vaccination**. This contains attenuated *Mycobacterium bovis* to stimulate the production of antibodies against both *M. bovis* and *M. tuberculosis*.

Cure Use of antibiotics such as streptomycin.

Disease file – malaria

OCR M2

Cause of disease

There are many variants of the malarial parasite, *Plasmodium* (protozoa).

Transmission of the microorganism

The vector which carries the *Plasmodium* is a female *Anopheles* mosquito. The mosquito feeds on a mammal which may be suffering from malaria. It does this at night by inserting its 'syringe-like' stylet into a blood vessel beneath the skin. The mosquito feeds on blood and digests the red blood cells which releases the malarial parasites. These burrow into the insect's stomach wall where they breed; some then move to the salivary glands. Next time the mosquito feeds it secretes saliva to prevent clotting of the blood. This secretion introduces the parasites into the person's blood, who is likely to contract the disease.

Malaria is endemic in the Middle East and Southern Asia, where the vectors of the disease live successfully. Global warming is beginning to have an effect on the distribution of the disease. New areas suitable as habitats for the *Anopheles* mosquito are appearing because of global warming.

Outline of the course of the disease and symptoms

After entry into the blood, **sporozoites** invade the **liver** releasing many **merozoites**. Each merozoites infects a **red blood cell** producing even more merozoites. Millions of these parasites are released into the blood causing a fever. As a result, the sufferer develops a range of symptoms including pains, exhaustion, aching, feeling cold, sweating and fever. The increased body temperature attracts mosquitoes even more, so a person with malaria acts as a reservoir for parasites.

Prevention

The most effective methods of prevention are those which **destroy the vector**. Spraying **insecticide** onto lake surfaces kills mosquito larvae.

Oil poured on lake surfaces prevents air entering the breathing tubes of the mosquitoes, so they die. Fish can be introduced into lakes as **predators** to eat the larvae. This is an example of **biological control**.

Combinations of these tests are used in different countries. Where there are outbreaks of the disease the systems are activated.

Sometimes ponds are drained to remove the mosquitoes' breeding area. People in areas where malaria is endemic cover up all waste tin cans and plastic containers. If they were to fill up with rain water then the mosquitoes have another habitat to breed in. The bacterium, *Bacillus thuringiensis* is used to destroy mosquitoes. Mosquito nets exclude mosquitoes from buildings and are even used over beds. Electronic insect killer techniques can be used which attract the mosquito via ultra-violet light then kill them by application of voltage. Drugs are used so that even if a person is bitten by a mosquito any *Plasmodia* entering the blood fail to develop further.

Cure

It is necessary to isolate and treat the sufferer. This also reduces the spread of the disease. Drugs are used to kill the parasites in the blood and reduce the symptoms. People are constantly attempting to find different ways of preventing this killer disease.

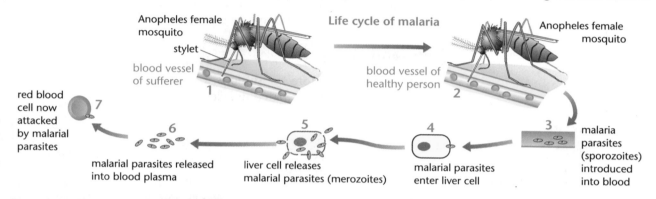

Progress check

A mosquito carries the malarial parasite, *Plasmodium*. The female mosquito feeds on a mammal by inserting its 'syringe-like' stylet into a blood vessel beneath the skin. The mosquito feeds on blood and digests the red blood cells releasing the malarial parasites. These burrow into the insect's stomach wall and breed there, then some move to the salivary glands. Next time the mosquito feeds it secretes saliva. The saliva introduces the parasites into the person's blood.

(a) (i) Which species of mosquito transmits malaria?
(ii) Which organism causes the disease, malaria?
(iii) Does every mosquito bite transmit malaria?

Give a reason for your answer.

(b) Suggest how to reduce the spread of malaria.

(a) (i) *Anopheles* (ii) *plasmodium* (iii) no – mosquito must feed on sufferer first.
(b) Drain ponds where the mosquitoes breed; kill the mosquitoes with insecticide; pour oil on ponds to kill the larvae; introduce insectivorous fish as a form of biological control; use drugs such as chloroquine to cure people suffering from the disease; isolate people suffering from the disease; spray mosquitoes with a suspension of *Bacillus thuringiensis*.

Disease file – AIDS (Acquired Immune Deficiency Syndrome)

OCR M2

Cause of disease

This is by HIV (human immune deficiency virus). It is a retrovirus, which is able to make DNA with the help of its own core of RNA.

Transmission of microorganism

This takes place by the exchange of body fluids, transfusion of contaminated blood, or via syringe needle 'sharing' in drug practices.

Scientists are constantly trying to find a **cure**. None has been found yet.

Outline of the course of the disease and symptoms

Destruction of T-lymphocyte cells

The HIV protein coat attaches to protein in the plasma membrane of a T-lymphocyte. The virus protein coat fuses with the cell membrane releasing RNA and reverse transcriptase into the cell. This enzyme causes the cell to produce DNA from the viral RNA. This DNA enters the nucleus of the T-lymphocyte and is incorporated into the host cell chromosomes. The gene representing the HIV virus is permanently in the nucleus from now on and can be dormant for years. It may become activated by an infection. Viral protein and viral RNA are made as a result of the infection.

Many RNA viral cores now leave the cell and protein coats are assembled from degenerating plasma membranes. Other T-lymphocytes are attacked. Cells of the lymph nodes and spleen are also destroyed. Viruses appear in the blood, tears, saliva, semen and vaginal fluids. The immune system becomes so weak that many diseases can now successfully invade the weakened body.

Prevention

Screening of blood before transfusions. Use of condoms and remaining with one partner. No use of contaminated needles.

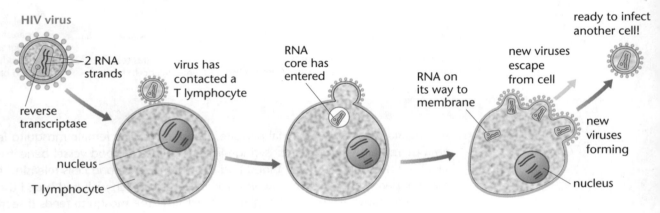

9.3 Immunity

After studying this section you should be able to:

- *describe and explain the action of the body's immune system*

Survival against the attack of pathogens

AQA A	M2
EDEXCEL	M3
OCR	M2

Many pathogenic organisms attack people. They are not all successful in causing disease. We have immunity to a disease when we are able to resist infection. The body has a range of ways to prevent the disease-causing organism from becoming established.

- A tough protein called **keratin** helps skin cells to be a formidable **barrier** to prevent pathogens entering the body.
- An enzyme, **lysozyme**, destroys some microorganisms and can be found in sebum, tears and saliva.
- **Hydrochloric acid** in the stomach kills some microorganisms.

- The bronchial tubes of the lungs are lined with cilia. Microorganisms which enter the respiratory system are often trapped in mucus which is then moved to the oesophagus. From here they move to the stomach where many are destroyed by hydrochloric acid or digested.
- Blood clotting in response to external damage prevents entry of microorganisms from the external environment.

The ways in which the body is adapted to prevent microorganisms entering the bloodstream are sometimes unsuccessful. When the microorganisms invade, then breed in high numbers, we develop the symptoms. White blood cells enable us to destroy invading microorganisms. They may destroy the microorganisms quickly before they have any chance of becoming established, so the person would not develop any symptoms. Sometimes there are so many microorganisms attacking that the white blood cells cannot destroy all of them. Once the pathogens are established the symptoms of a disease follow, but for most diseases, after some time, the white blood cells eventually overcome the disease-causing organisms.

The roles of the white blood cells (leucocytes)

There are a number of different types of leucocytes. They are all produced from stem cells in the bone marrow. Different stem cells follow alternative maturation procedures to produce a range of leucocytes. Leucocytes have the ability to recognise self chemicals and non-self. Only where non-self chemicals are recognised will a leucocyte respond. Proteins and polysaccharides are typical of the complex molecules which can trigger an immune response.

Phagocytes

Phagocytes can move to a site of infection through capillaries, tissue fluid and lymph as well as being found in the plasma. They move towards pathogens which they destroy by the process of phagocytosis. This is often called engulfment and involves the surrounding of a pathogen by pseudopodia to form a food vacuole. Hydrolytic enzymes complete the destruction of the pathogen.

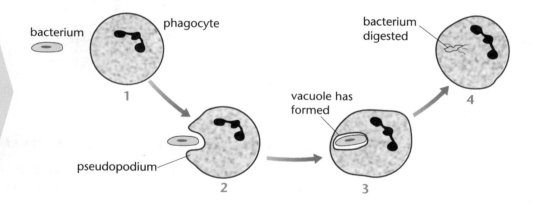

What is an antigen?

AQA A   M2
EDEXCEL   M3
OCR   M2

As an individual grows and develops, complex substances such as proteins and polysaccharides are used to form cellular structures. Leucocytes identify these substances in the body as 'self' substances. They are ignored as the leucocytes encounter them daily. 'Non-self substances', e.g. foreign proteins which enter the body, are immediately identified as 'non-self.' These are known as antigens and trigger an immune response.

> White blood cells (leucocytes) constantly check out proteins around the body. Foreign protein is identified and attack is stimulated.

Lymphocytes

There are two types of lymphocyte, **B-lymphocytes** and **T-lymphocytes**.

B-lymphocytes begin development and mature in the bone marrow. They produce antibodies, known as the **humoral response**.

T-lymphocytes work alongside phagocytes known as macrophages; this is known as the **cell-mediated response**. A macrophage engulfs an antigen. This antigen remains on the surface of the macrophage. T-lymphocytes respond to the antigen, dividing by mitosis to form a range of different types of T-lymphocyte cells.

- **Killer T-lymphocytes** adhere to the pathogen, secrete a toxin and destroy it.
- **Helper T-lymphocytes** stimulate the production of antibodies.
- **Suppressor T-lymphocytes** are inhibitors of the T-lymphocytes and plasma cells. Just weeks after the initial infection, they shut down the immune response when it is no longer needed.
- **Memory T-lymphocytes** respond to an antigen previously experienced. They are able to destroy the same pathogen before symptoms appear.

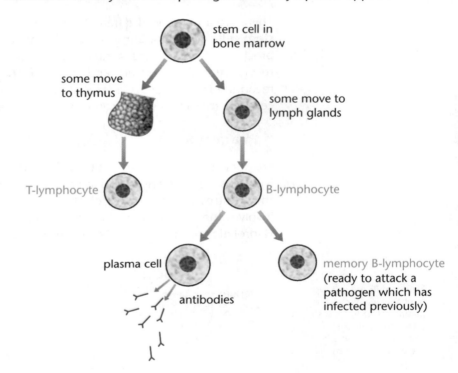

How do antibodies destroy pathogens?

The diagram above shows antibodies binding to antigens. The descriptions below show what can happen immediately after the binding takes place.

There are four main ways in which antibodies destroy pathogens.

- **Precipitation**, by linking many antigens together. This enables the phagocytes to engulf them.
- **Lysis**, where the cell membrane breaks open, killing the cell.
- **Neutralisation** of a chemical released by the pathogen, so that the chemical is no longer toxic.
- Attachment of **opsonins** to the membrane of the microorganism, which causes them to clump together. They then attach each pathogen to a phagocyte which engulfs them.

Sample question

The graph below shows the relative numbers of antibodies in a person's blood after the vaccination of attenuated viruses. Vaccinations were given on day 1 then 200 days later.

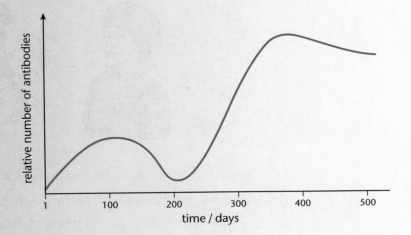

In examinations you are regularly given graphs. Make sure that you can link the idea being tested. This should help you recall all of the important concepts needed. All you need to do after this is **apply** your knowledge to the given data.

(a) Why is it important that viruses used in vaccinations are attenuated? [1]

If they were active then the person would contract the disease.

(b) Suggest **two** advantages of giving the second vaccination. [2]

A greater number of antibodies were produced.
The antibodies remain for much longer after the second vaccination.

(c) Which cells produced the antibodies during the primary response? [1]

B-lymphocytes.

(d) Why was there no delay in the secondary response to vaccination? [3]

Because the first vaccination had already been given, memory B-lymphocytes had been produced which respond to the viruses more quickly.

(e) Describe how a virus stimulates the production of antibodies? [3]

Antigen in the protein 'coat' or capsomere stimulate the B-lymphocytes.

(f) Apart from producing antibodies, outline FOUR different ways that the body uses to destroy microorganisms. [4]

If you gave B-lymphocytes as a response it would be wrong! B-lymphocytes secrete antibodies.

Phagocytes by engulfment; T-lymphocytes attach to microorganism and destroy them; hydrochloric acid in the stomach; lysozyme in tears.

Practice examination questions

1 The diagram shows how the bacterium which causes typhoid can be transmitted from one person to another.

(a) Name the method of disease transmission shown in the diagram. [1]

(b) Sometimes the bacteria infect people but they do not develop symptoms.

   (i) What term is given to this group of people? [1]

   (ii) Explain why these people may be a greater danger to a community than those who actually suffer from the disease. [2]

(c) (i) What can be given to a person infected with typhoid to help destroy the bacteria? [1]

   (ii) Explain the role of each of the following in destroying typhoid bacteria.
   Phagocyte
   B-lymphocyte
   T-lymphocyte [6]

2 The diagram shows the response of B-lymphocytes to a specific antigen.

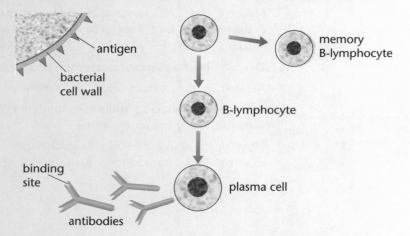

(a) (i) A plasma cell is bigger than a B-lymphocyte.
      Suggest an advantage of this. [1]

   (ii) Describe the precise role of antibodies in the immune response. [3]

   (iii) What is the advantage of memory B-lymphocytes? [2]

(b) What is an auto-immune disease? Give an example. [2]

(c) A person contracts the virus which causes the common cold.
    Suggest why their lymphocytes fail to destroy the pathogen. [1]

Practice examination answers

Chapter 1 Biological molecules

1 (a) $\dfrac{90}{100} = 0.9$ [2]

(b) X [1]

(c) The solvent front would have reached the edge of the paper. [1]

[Total: 4]

2 (a)
RCOOH   HOCH$_2$
           |
RCOOH + HOCH
           |
RCOOH   HOCH$_2$
fatty acids   glycerol [2]

(b) Emulsion test: add the sample to ethanol and mix; decant or pour into water; if a fat is present a white emulsion forms on the surface. [3]

[Total: 5]

3 (a) peptide bond/peptide link [1]

(b) –COOH/carboxylic acid [1]

(c) primary structure; amino acids in a chain [2]

[Total: 4]

4 (a) The latent heat of evaporation is large so lots of energy is needed to evaporate water/in sweating, much body heat is needed for evaporation. [2]

(b) High specific heat capacity means that the water needs a lot of heat energy to increase temperature significantly, therefore an organism will not overheat easily. [2]

(c) Cohesive forces aid the movement of water up the xylem. [2]

[Total: 6]

Chapter 2 Cells

1 (a) phospholipid [1]

(b) (i) Substance approaches a carrier protein molecule; carrier protein activated by ATP; protein changes shape allowing the substance into the cell.

(ii) Substance approaches a carrier protein; this may be a channel protein and substances pass through without any ATP necessary. [4]

[Total: 5]

2 (a)

	diffusion	facilitated diffusion	active transport
molecules move from where they are in high concentration to low concentration	✓	✓	
molecules move from where they are in low concentration to high concentration		✓	✓
a protein carrier is needed			✓

[3]

(b) phagocytosis/endocytosis/exocytosis/osmosis [any two for 2 marks] [2]

[Total: 5]

3 Cuticle reflects some light so that the leaf does not have excess, which would dehydrate the leaf. [1] Palisade cells have most chloroplasts so that they can capture maximum light. [1] The chloroplasts are motile, able to move in the cytoplasm to absorb most light. [1] Large air spaces in the spongy mesophyll store carbon dioxide for photosynthesis. [1] Stomata open to allow high amounts of carbon dioxide to enter the leaf. [1] Water is supplied to photosynthetic cells by the xylem. [1] Glucose is rapidly removed by the phloem. [1] [7]

[Total: 7]

4 (a) cell wall (not cellulose); no true nucleus; no mitochondria; plasmids [any two for 2 marks] [2]

(b) mitochondria; nucleus; Golgi body; large ribosomes/rough endoplasmic reticulum [any two for 2 marks] [2]

[Total: 4]

5 (a) (i) total number of yeast cells in 10 squares = 76
$\dfrac{76}{10}$ = 7.6 (average number per square) [2]

(ii) volume of one square = (0.0001) mm^3
number of cells in one square = 7.6
number in 1 cm^3
$= \dfrac{1000 \times 10 \text{ (dilution factor)} \times 7.6}{0.00001}$
$= 7.6 \times 10^9$ [3]

(b) so that the suspension is homogeneous [1]

[Total 6]

6 (a) lens X = objective; lens Y = condenser [2]

(b) electrons would collide with air molecules [1]

(c) (i) artefact [1]

(ii) ignore the artefact, because it is not part of normal structure and is only present due to preparation of the specimen. [1]

[Total 5]

Chapter 3 Enzymes

1 (a) The urea molecules bind with receptor molecules on the biological recognition layer; the transducer measures and amplifies a signal; electrical signal identifies urea. [3]
   (b) Level of voltage. [1]
   [Total: 4]

2 (a) lock and key – the substrate is a similar shape to the active site; it fits in and binds with the active site like a key (substrate) fitting into a lock (active site); induced fit – the substrate is not a matching 'fit' for the active site, but as the substrate approaches, the active site changes into an appropriate shape.
   (b) Endopeptidases break down peptide links in the middle parts of polypeptides; exopeptidases break down the peptides links at the ends of polypeptides, removing the outer amino acids. [4]
   [Total: 4]

3 They do not contaminate the product; they can be used again and again. [2]
   [Total: 2]

4 • The tertiary structure of the enzyme is responsible for the further folding of the protein;
   • this gives the shape of the active site;
   • the active site in amylase is specific to starch, lipid is unable to bind to the active site of amylase, therefore amylase cannot break down lipid. [3]

5 (a) starch [1]
   (b) thermostable enzymes are effective at high temperatures [1]
   [Total: 2]

6 (a) Stain removal directly proportional to protease concentration. [2]
   (b) (i) At 6 units of protease dm^{-3} the maximum rate is reached; after this amount there was no further increase in stain removal. [1]
      (ii) Protein molecules bind to the enzyme's active site; water molecules are part of the hydrolysis mechanism to break down the protein; polypeptides or amino acids are produced which readily move from the cloth. [3]
      (iii) so that a lot of different proteins can be broken down by the same enzyme. [1]
      (iv) polypeptide/amino acid [1]
   (c) lipase; carbohydrase/amylase [2]
   [Total: 9]

7 (a) The substrate molecule collides with the active site of the enzyme; as it approaches, the active site changes shape to become compatible with the substrate shape. [2]
   (b) Non-competitive inhibitor molecule binds with part of enzyme other than the active site; as a result the active site changes shape; so the substrate can no longer bind with the active site. [3]
   [Total: 5]

8 (a) NH_2 (amino group) [1]
   (b) Endopeptidases break down the middle of polypeptides; exopeptidases break down the outside peptide bonds to release amino acids; every time the endopeptidase breaks the middle of the polypeptide it reveals two more ends for the exopeptidase to work on. [3]
   [Total: 4]

Chapter 4 Exchange

1 (a) (i) No change in size because the water potential inside the cell equals the water potential of the solution outside the cell. [1]
      (ii) The water potential of the solution outside the cell is more negative than the water potential inside the cell. [1]
      (iii) The water potential of the solution outside the cell is less negative than the water potential inside the cell. [1]
   (b) osmosis [1]
   [Total: 4]

2 (a) They both use a protein carrier molecule. [1]
   (b) Active transport needs energy or mitochondria, whereas facilitated diffusion does not.
      OR Active transport allows molecules to move from a lower concentrated solution to a higher concentrated solution.
      OR Active transport allows molecules to move against a concentration gradient. [1]
   [Total: 2]

3 (a) • a flat, thin blade allows maximum light absorption;
      • the leaf has a waxy cuticle to reflect excess light but allow entry of enough light for photosynthesis;
      • the mesophyll cells have many chloroplasts to absorb the maximum amount of light;
      • the palisade cells pack closely together to absorb the maximum amount of light;
      • guard cells open stomata to allow carbon dioxide in and oxygen out during photosynthesis;
      • air spaces in the mesophyll store lots of carbon dioxide for photosynthesis;
      • xylem of the vascular bundles bring water to the leaf for photosynthesis;
      • phloem takes the carbohydrate away from the leaf after photosynthesis. [any 6 points] [6]
   (b) thick waxy cuticle; low number of stomata; hairs on epidermis (which reduce turbulence). [3]
   [Total: 9]

4 (a) alveoli have a very high surface area; they are very close to many capillaries; capillaries are one cell thick/very thin/have squamous epithelia; they are kept damp which facilitates diffusion. [4]
   (b) • there are many gill filaments, which give a very

high surface area;
• the gill filaments are very thin;
• they have many capillaries;
• they have gill plates which increase the surface area further;

• the blood and water directions are opposite which maximises diffusion/they use countercurrent flow to maximise diffusion. [any 4 points] [4]
[Total: 8]

Chapter 5 Transport

1 (a) to the body core [1]
(b) less blood reaches the superficial capillaries of the skin; so less heat is lost by conduction, convection and radiation; blood in body core better insulated by the adipose layer of the skin [4]
[Total: 5]

2 (a) 60 x 120 x 5 = 36 000 ml / 36 litres [2]
(b) • blood is transported more quickly;
• more oxygen taken up at the lungs/more carbon dioxide excreted at the lungs;
• more oxygen reaches the muscles;
• more glucose reaches the muscles;
• so muscles contract more effectively.
[any 4 points] [4]
[Total: 4]
(c) (i) • slower breathing rate;
• more alveoli accessed for exchange;
• intercostal muscles more effective.
[any 2 points] [2]
(ii)• improved muscle tone;
• greater muscular strength;
• more capillaries in muscles. [any 2 points] [2]
[Total: 4]

3 (a) Lung cancer – cause: tobacco or tars are carcinogenic; mutates the lung cells
– symptom: malignant growths/abnormal growth. [2]
(b) Bronchitis – cause: smoking/damp cold conditions; inflammation of the bronchi or bronchioles – symptoms: coughing/phlegm/increased chance of pneumonia. [2]
(c) Emphysema – cause: smoking; damaged alveoli/walls between alveoli reduce, so surface area for gaseous exchange less – symptoms: breathless, cannot obtain enough oxygen. [2]
[Total: 6]

4 (a) The amount of water lost by transpiration is exactly matched by the amount taken up by the leaf. [1]
(b) (i) xylem [1]
(ii) fill the potometer under water; operate the valve to get rid of air bubbles; when removing the leaf from the tree the stalk or petiole must be put in water immediately. [2]
(c) Volume of water = πr^2 x 32 mm x 60
= $\frac{22}{7}$ x 1 x 1 x 32 x 60
= 6034.3 mm^3 [3]
[Total: 7]

5 • extensive root system to absorb maximum water;
• large amount of water storage in leaves;
• thick cuticle;
• low numbers of stomata/sunken stomata;
• hairs to reduce turbulence. [any 3 points] [3]
[Total: 3]

6 (a) SAN/Sinoatrial node [1]
(b) (i) slows heart rate [1]
(ii) speeds up the rate [1]
(iii) speeds up the rate [1]
[Total: 4]

7 (a) (i) sieve tube [1]
(ii) companion cell has nucleus plus ribosomes; which make the proteins or enzymes; supplies enzymes to sieve tube via plasmodesmata. [3]
(b) (i) It is an active process; a pump is involved; phloem contents under high pressure. [3]
(ii) no starch in the phloem contents; did not change to blue-black; contained reducing sugar; did change to brick red. [4]
(iii) Hot wax kills phloem cells; so they cannot transport the radioactive carbohydrate; transport by phloem is an active process. [3]
[Total: 14]

Chapter 6 Genetic code

1 (a) 5; There were 5 cuts along the piece of DNA. So the base sequence binding to the active site of the enzyme occurred 5 times. [2]
(b) sticky end [1]
(c) 48502 – 44123 = 4379 [1 mark for working and 1 mark for correct answer.] [2]
(d) They cut the DNA into pieces; electrophoresis then used/voltage applied; different DNA samples compared and from same person all DNA samples will form a pattern like a bar-code. [8]
[Total: 3]

2 (a) metaphase [1]
(b) 2n [2]
[Total: 3]

3 (a) The gene is cut out from the human DNA using a restriction endonuclease. [1] A bacterial plasmid is cut using the same restriction endonuclease. [1] The human DNA is incorporated into the plasmid; [1] with ligase. [1] The plasmid now returned to bacterium. [1] Bacterium clones the plasmid. [1] [6]
(b) Sterilise the fermenter to kill contaminant microorganisms; supply nutrients plus the transgenic bacteria; use paddle wheel to ensure nutrients plus microorganisms make contact; adjust the pH during process/neutralise; temperature sensor plus water jacket to keep reaction at optimum; add air via air filter to remove contaminant microorganisms. [5]
(c) insulin [1]
[Total: 12]

Practice examination answers

4 (a) adenine and thymine are similar proportions because adenine binds with thymine; cytosine and guanine are similar proportions because cytosine binds with guanine [2]

(b) They should be identical in number but the scientists were operating at the limits of instrumentation. [1]

(c) Organic bases form the codes for different amino acids. Different sequences of amino acids form the different proteins specific to a species. [2]
[Total: 5]

5 (a) restriction endonuclease [1]

(b) Yes, the egg DNA shares several common bonds with the parents DNA. [1]

(c) checking out who is the father of a child/paternity cases; crimes where blood samples or tissue or saliva is left and checked against suspects [2]
[Total: 5]

6 (a) Identify the specific section of DNA which contains the gene; this can be done using reverse transcriptase; insert DNA into a vector/insert into *Agrobacterium tumefaciens*; this bacterium/this vector then passes the DNA into the recipient cell. [5]

(b) herbicide kills weeds; which reduces competition; for light or water or minerals; soya plants unharmed [3]

(c) the ability to resist the effect of herbicide could transfer to weeds; so the herbicide no longer effective on weeds; resistant weeds spread into field [3]

(d) fear that the new gene will be passed to other plants by interbreeding; fear that potentially toxic chemicals may be consumed; fear that the beans have not been tested enough [2]
[Total: 13]

Chapter 7 Continuity of life

1 (a) (i) corpus luteum [1]
(ii) via the blood [1]

(b) no, because the progesterone level fell [1]

(c) endometrium would detach and miscarriage take place [1]
[Total: 4]

2 (a) (i) this causes the ovules to ripen before the pollen [1]
(ii) this causes the pollen to ripen before the ovules [1]

(b) (i) no need to have more than one tree [1]
(ii) do not have hybrid vigour; greater chance of recessive disadvantageous character in offspring [2]
[Total: 5]

3 (a) (i) insect, because of the rough outside which sticks to insects or the converse, because it is not aerodynamic [1]
(ii) tube nucleus or tube growth [1]

(b) one male nucleus fuses with the egg cell; to form the embryo; one male nucleus fuses with the polar nuclei; to form the endosperm [4]
[Total: 6]

Chapter 8 Energy and ecosystems

1 (a) (i) When given fertiliser the grasses competed for resources better that the legumes; some legume species could not grow in these conditions. [2]
(ii) Without fertiliser the grass species did not have enough minerals so did not compete as well; the legumes fixed nitrogen in root nodules so could grow effectively. [2]

(b) Cows grazed on some species more than others/perhaps trampling by cattle destroyed some species but others were tougher and survived/perhaps waste encouraged the growth of some species whereas others were destroyed. [1]
[Total: 5]

2 (a) electronic systems locate shoals of fish accurately; very large nets/small-mesh nets [2]

(b) many of the breeding size fish have already been caught; some fish may never reach breeding size as they are caught before they reach this size [2]

(c) • trawling destroys some invertebrates;
• they may be the food of other organisms in a food web, so some animals may die out as a result;
• overfishing reduces fish numbers so that their predators may ultimately die out;
• nets catch other than target fish in the nets.
[any 3 points] [3]

(d) agree to quota numbers of fish; exclusion zones/exclusion times [2]
[Total: 9]

3 (a) (i) pioneer or primary coloniser [1]
(ii)
• algae cut off light from plants underneath;
• they die as a result;
• bacteria or fungi or saprobiotics decay them;
• they use a lot of oxygen;
• fish die due to not enough oxygen;
• blood worms increase in number as they are adapted to small amounts of oxygen.
[any 5 points] [5]

(b) • marginal plants or irises were introduced;
• they spread;
• each year the foliage died and rotted;
• this organic material or humus added to the soil or mud;
• secondary colonisers spread from other areas;
• succession took place.
[any 4 points) [4]
[Total: 10]

4 (a) more fish escape through the bigger holes; and go on to breed [2]

(b) legislation/regulations/rules/penalties/laws [1]

(c) its predators may increase correspondingly; the